SPACEFARERS

Images of Astronauts and
Cosmonauts in the Heroic Era of
SPACEFLIGHT

Edited by
MICHAEL J. NEUFELD

A SMITHSONIAN CONTRIBUTION TO KNOWLEDGE

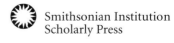
Smithsonian Institution
Scholarly Press

WASHINGTON, D.C.
2013

Published by SMITHSONIAN INSTITUTION SCHOLARLY PRESS
P.O. Box 37012, MRC 957
Washington, D.C. 20013-7012
www.scholarlypress.si.edu

Compilation copyright © 2013 by Smithsonian Institution

Text of Chapters 1, 5, and 7 is in the public domain. The rights to all other text and images in this publication, including cover and interior designs are owned by the Smithsonian Institution, contributing authors, or third parties and may not be reproduced, stored in a retrieval system, or transmitted in any form or by any means, electronic, mechanical, photocopying, recording, or otherwise, without the prior permission of the publisher.

Dust jacket image: *Opening the Space Frontier, The Next Giant Step*, mural by Robert T. McCall in the NASA Johnson Space Center, Houston, Texas, 1979. Courtesy McCall Studios, http://www.mccallstudios.com/.

Library of Congress Cataloging-in-Publication Data:

Spacefarers : images of astronauts and cosmonauts in the heroic era of spaceflight / edited by Michael J. Neufeld.
     p. cm. — (A smithsonian contribution to knowledge)
  Includes bibliographical references and index.
  ISBN 978-1-935623-19-9 (alk. paper)
 1. Astronautics in mass media. 2. Astronautics—Social aspects. 3. Astronautics and state. 4. Astronauts—United States. 5. Astronauts—Soviet Union. I. Neufeld, Michael J., 1951– II. Series: Smithsonian contribution to knowledge.
  P96.A792S63 2013
  629.45—dc23
                                                           2013000432

Printed in the United States of America

∞ The paper used in this publication meets the minimum requirements of the American National Standard for Permanence of Paper for Printed Library Materials Z39.48–1992.

CONTENTS

INTRODUCTION — 1
Michael J. Neufeld

CHAPTER 1 — 9
Setting the Scene for Human Spaceflights:
Men into Space and *The Man and the Challenge*
Margaret A. Weitekamp

CHAPTER 2 — 35
"Capsules Are Swallowed":
The Mythology of the Pilot in American Spaceflight
Matthew H. Hersch

CHAPTER 3 — 57
Nostalgia for the Right Stuff:
Astronauts and Public Anxiety about a Changing Nation
James Spiller

CHAPTER 4 — 81
The Fiftieth Jubilee:
Yuri Gagarin in the Soviet and Post-Soviet Imagination
Andrew Jenks

CHAPTER 5 — 107
Astronauts and Cosmonauts into Frenchmen:
Understanding Space Travel through the Popular Weekly *Paris Match*
Guillaume de Syon

CHAPTER 6 — 125
They May Remake Our Image of Mankind:
Representations of Cosmonauts and Astronauts in Soviet
and American Space Propaganda Magazines, 1961–1981
Trevor S. Rockwell

CHAPTER 7 — 149
Bringing Spaceflight Down to Earth:
Astronauts and The IMAX *Experience*®
Valerie Neal

CHAPTER 8 — 175
You've Come a Long Way, Maybe:
The First Six Women Astronauts and the Media
Jennifer Ross-Nazzal

CHAPTER 9 — 203
Warriors and Worriers:
Risk, Masculinity, and the Anxiety of Individuality
in the Literature of American Spaceflight
Margaret Lazarus Dean

ACKNOWLEDGMENTS	227
SELECTED BIBLIOGRAPHY	229
CONTRIBUTORS	239
INDEX	243

Introduction
Michael J. Neufeld

This volume had its origins in a conference held at the headquarters of the National Aeronautics and Space Administration (NASA) in Washington, D.C., in April 2011. The conference "1961/1981: Key Moments in Human Spaceflight" commemorated the fiftieth anniversary of the first human spaceflights by the Soviet Union and the United States, and the thirtieth anniversary of the first space shuttle launch, which took place exactly twenty years after Yuri Gagarin's pioneering orbit on April 12, 1961. This book, however, does not present most of the papers that were given, nor are all the chapters based on papers read at the conference. Rather, in consultation with the NASA History Program Office, I decided that the one coherent theme that emerged from several papers was the image of astronauts and cosmonauts in the media, state propaganda, and popular culture. Social and cultural history of the first three decades of human spaceflight was prominent at the conference, reflecting the maturation of the space history subdiscipline and its growth beyond the traditional technical, programmatic, and political history narratives. Stephen Garber of NASA History and I therefore asked a couple of the participants to revise their papers or submit new papers to fit the theme of the images of astronauts and cosmonauts, and we solicited one new article to cover at least one theme that had been relatively neglected, the women astronauts of the shuttle program.

Marie Lathers, the author of *Space Oddities: Women and Outer Space in Popular Film and Culture, 1960–2000*, has asserted that for the United States (and one might add for the rest of the world), 1961 was "a fundamental rupture . . . , when real men went into orbit."[1] Two millennia of imaginary space voyagers and a century of more realistic depictions of travelers in science fiction and spaceflight advocacy were suddenly displaced by the existence of actual spacefarers and their adventures. The first cosmonauts and astronauts to fly, almost all fighter pilots, became iconic figures, with images drawn from several archetypes, such as heroic aviators and explorers. In the Soviet Union, because of obsessive secrecy, that fame appeared at the moment when their launch was announced; in the United States, celebrity began two years before the first flights, with the introduction of the seven Mercury astronauts to the media. Both corps of spacefarers were represented in their respective media and government propaganda as shining examples of masculinity and patriotism representing all the best values of their competing cold war systems.

The missions of the first astronauts and cosmonauts opened a period I call "the heroic era" of human spaceflight—both for the media representation of their image and the actual danger of their occupation.[2] Seven of them were killed in ground tests or spaceflights between 1967 and 1971, and others died in plane crashes and training accidents throughout the sixties. The superpower race for firsts, culminating in the Apollo 11 Moon landing in July 1969, set the agenda and increased the risk. The United States built one-man Mercury, two-man Gemini, and three-man Apollo spacecraft, plus a two-man lunar lander; the Soviet Union created the one-cosmonaut Vostok capsule, modified it as the two- or three-passenger Voskhod, then built a Soyuz spacecraft that could carry up to three. Soviet engineers modified the Soyuz to loop around and orbit the Moon, plus a one-man lander to get to the surface, but no Soviet cosmonaut ever flew them. In the aftermath of the American triumph, the Soviet Union denied its lunar plans and shifted to space station missions, whereas the United States, after six landings, retreated to low Earth orbit. It used Apollo spacecraft to get to one temporary station, Skylab, and to link up with a Soviet Soyuz in 1975, during a brief period of détente. In 1972, NASA had been given the go-ahead to build a reusable space shuttle, carrying up to seven or eight astronauts, in order to make Earth orbit missions and a permanent space station easier, but it did not fly until 1981.

When the heroic era ended is arguable. Two shuttle accidents in 1986 and 2003 demonstrate that human spaceflight has never become routine. However, for the purposes of this book we will define it as ending with the *Challenger* disaster of 1986 (although two chapters do look beyond that point). After *Challenger,* space travel resumed its trajectory toward becoming frequent and anonymous; the end of the cold war in 1989 resulted in the demise of the Soviet Union and the rise of cooperative rather than competitive human space programs (the Chinese excepted).

As human spaceflight objectives became more ambitious and spacecraft technology improved, scientists and women (and women scientists) were added to the ranks of spacefarers, beginning with the flight of Valentina Tereshkova in 1963, although it would be nineteen years before another woman flew. That changing crew composition, especially in the 1980s, inevitably complicated the masculine pilot-hero image that had made such a deep impression on popular culture, East and West, during the early space race. The ordinary human peccadilloes of the icons, drinking, womanizing, etc., also began to intrude on the image of perfection. In the Soviet Union, moreover, cosmonauts found themselves caught in contradictions. Being dishonest or evasive was mandatory in order to protect secrecy and uphold the official line that everything was going exactly as planned, when manifestly it was not.[3] Dissent and satire regarding the space program in the Eastern Bloc, however, had to be confined to jokes and the work of unofficial artists. In the West, by contrast, a free press and media could create alternative or unflattering images of astronauts, for example, in Hollywood movies, especially as the cold war consensus eroded. Nonetheless, the heroic vision of spacefarers remained dominant in the media, even when few in the public knew their names anymore.

Because the topic is fairly new, the existing scholarly literature on images of astronauts and cosmonauts has only developed rather recently. Historians and cultural scholars of the USSR and Russia have led the way, not because, as many American historians like to think, it is so hard to research the actual archives of the Soviet space program, but because of the flourishing of cultural and social history of the Soviet Union. Coming out of a vibrant interest in the cultural and propaganda dimensions of Stalinism, Eastern European history scholars have turned their attention to the Nikita Khrushchev "thaw" period, in which the early spectacular space successes induced a "cosmic enthusiasm" in the general public that was not just the product of official activity. Two edited collections have recently appeared on this topic; each contains articles on cosmonauts as icons.[4]

On American spacefarers, scholars have employed interdisciplinary approaches from history to gender and film studies, but more attention has been paid to fictional characters than actual astronauts. A case in point is Lathers' very stimulating feminist analysis of women in space since 1960, which deals almost exclusively with movies and essentially avoids the topic of real women astronauts after their first selection in 1978. One of the few articles about astronauts as icons, until recently, was written by Roger Launius. Although it is nominally about the Moon program, he begins with the centrality of the Mercury 7 experience, which shaped how all astronauts were perceived. He sees five essential components to their "myth": the astronaut as "everyman," "defender of the nation," "fun-loving young man," "virile, masculine representative of the American ideal," and "hero."[5]

Tom Wolfe elaborated that myth in his "nonfiction novel" *The Right Stuff*, published at the end of the 1970s.[6] Wolfe's relentlessly popular account helped break the conspiracy of silence around the astronauts' personal behavior, but the net effect of his book was to bolster the received image. A more sophisticated social and cultural history of the human space program in the United States did not emerge until the end of the 1990s.[7] Before then, the primary sources on astronauts remained their numerous, often ghost-written memoirs. But in the last few years several scholars have turned their attention to cultural representation of American professional space travelers, and many of them are contributors to this volume, notably Matthew H. Hersch, who published *Inventing the American Astronaut* just as this book was going to press.[8]

The nine chapters in this book fall into three sets of three. The first trio addresses the origins of American astronaut images and their elaboration over time. In the first chapter Margaret A. Weitekamp examines two U.S. television programs that were broadcast for one season, 1959–1960, *Men into Space* and *The Man and the Challenge*. Although not about the real astronauts, who were introduced to the public on April 9, 1959, when these shows were already in production, they illuminate the exact moment of change, when American television and movie producers transitioned from the traditional "space opera" (*Buck Rogers, Captain Video*, etc.) to dramas in which spacefarers are imagined as actual U.S. officers and scientists. Many of the rapidly emerging stereotypes are incorporated into those two programs, notably the prominent representations of masculinity that Weitekamp analyzes. What was not so clearly expressed was the test-pilot image of the astronaut that materialized with the Mercury 7. Matthew H. Hersch, in Chapter 2, addresses that stereotype, where it came from, and how it shaped the astronaut corps and the popular media's perception of the astronaut. In Hersch's analysis, the pilot "mythology" was so dominant that scientist-astronauts were marginalized or confined to secondary roles at NASA, whereas the popular media could depict competent spacefarers in no other way than as pilots. In Chapter 3, James Spiller analyzes frontier images of the astronauts. As Weitekamp and Hersch had already noted, the Western frontier already loomed large in the American imagination of space travel. Spiller compares the Mercury 7 to the early shuttle astronauts of the 1980s as they were discussed by NASA and the American print media. He finds that the former were represented as rugged individualist frontier heroes reminiscent of the rhetoric of President Theodore Roosevelt, whereas the shuttle astronauts were described as pioneer settlers in the manner of the historian Frederick Jackson Turner. Both these frontier images betrayed the cultural, social, and political anxieties of the United States in a period of disorienting change.

The second trio of chapters looks at cosmonauts or cosmonauts and astronauts comparatively. Andrew L. Jenks, author of the first scholarly biography of Yuri Gagarin, comprehensively examines Gagarin in Soviet and Russian myth and memory.[9] Chosen for the first flight in part because he was the perfect embodiment of an ethnic Russian from modest circumstances rising to become a loyal servant of the state, Gagarin immediately struck an enormous chord with the Soviet public. The village of his birth and the town where he grew up became pilgrimage points, sustaining a tourism that propped up those places, notably since the collapse of the Soviet Union. That unexpected disaster would spur some metropolitan intellectuals to question the Gagarin cult or play with it artistically, but Jenks finds that in provincial Russia the original myth lives on, providing nostalgic and nationalistic comfort. In Chapter 5, Guillaume de Syon then examines cosmonaut and astronaut images in France, as represented by the dominant picture magazine, *Paris Match*. He finds it worked to personalize the alien technology of spaceflight and to make the people it featured into characters its French middle-class target audience could recognize. But soon after the first Moon landing, the magazine turned its attention elsewhere, a trend accelerated by the increasingly difficult market for the print glossy (as was true elsewhere). The only exception to the declining space coverage was a brief flurry around the voyages of the first French "spationauts" in the 1980s. Finally, Trevor S. Rockwell looks at how the United States and the USSR represented their respective spacefarers in their official illustrated magazines, *Amerika* and *Soviet Life*, which were distributed in each other's countries. Beyond the surface layer of political rhetoric, he finds that the two magazines depicted the astronauts and cosmonauts nearly identically, as cool professionals, peaceful space explorers, and warm family men (plus one woman, Valentina Tereshkova), who were bringing an inevitable future of progress in space and on the ground. It was a propaganda of "peaceful coexistence" with a common heritage in Enlightenment progress ideology, notwithstanding the stark contrasts in economics and politics.

The final trio of chapters primarily addresses the space shuttle astronauts. In Chapter 7, Valerie Neal considers how the giant-screen IMAX® movies, an innovation of the 1970s, shaped the perception of the astronauts in the 1980s and beyond. In particular, *The Dream Is Alive* (1985) was the first to include footage done on orbit, setting the standard for later films. The everyman (or everywoman) image prevailed in these movies, making the audience feel as they too were experiencing launch and weightlessness. Little explicit commentary was made about risk, and the work of the shuttle pilots was not elevated over the mission specialists, who were scientists, engineers, and physicians, often female. Neal's conclusions implicitly clash with Hersch's regarding the public image of astronauts in the shuttle era, but only further research can resolve whether the IMAX films, shown primarily in museums but to audiences in the millions, significantly eroded the dominant hero-pilot stereotype that still ruled Hollywood films. Jennifer Ross-Nazzal, in Chapter 8, then examines the media reception of the first six women astronauts chosen in 1978, uncovering a blatant and persistent sexism. They were confronted with an obsession with appearance and with a double standard regarding how to combine careers with homelife and motherhood. This attitude was consistent with American media attitudes at the time. It only seemed to have moderated when their flights in space became old news in 1985, perhaps only because it was old news. In the final chapter, Margaret Lazarus Dean, a novelist and literature scholar, compares the astronaut images in Wolfe's *The Right Stuff* to a contemporary novel, Stephen Harrigan's *Challenger Park* (2006), which includes a woman astronaut in Houston as one of the two main characters.[10] Dean returns to the theme of masculinity first discussed in this volume by Margaret Weitekamp and compares it to how femininity is depicted in *The Right Stuff*, through the astronauts' wives, and in *Challenger Park*, through its conflicted woman astronaut who cannot seem to reconcile her occupation with motherhood. Her conclusion is a little disturbing: even in the early twenty-first century, with spaceflights by women seemingly routine, we still are not completely accepting of the idea of the woman astronaut. The masculine hero-pilot may indeed still reign supreme.

We hope the essays in this book will stimulate further work in this field, in particular because of its careful exploration of the complex roles that gender, aviation, the frontier, film, nationalism, and cold war propaganda played in the social and cultural processing of the images of astronauts and cosmonauts. Clearly, there is much more to be done in examining the images of spacefarers globally, across time, and across a variety of media. Another challenge will be to situate those images in what Alexander C. T. Geppert has called modern "astroculture"—the whole complex of ideas and images related to space travel, astronomy, aliens, etc., that have now penetrated transnational popular and elite cultures.[11] Human spaceflight may or may not flourish in coming decades, but it seems certain that humans who travel in space will continue to be the object of a rich cultural discourse.

Notes

1. Marie Lathers, *Space Oddities: Women and Outer Space in Popular Film and Culture, 1960–2000* (New York: Continuum, 2010), 8.

2. The term and the concept is derived from Alex Roland's "Barnstorming in Space: The Rise and Fall of the Romantic Era of Spaceflight, 1957–1986," in *Space Policy Reconsidered*, edited by Radford Byerly, Jr. (Boulder, CO: Westview Press, 1989), 33–52. I have, however, chosen the adjective heroic as it addresses the image of early spacefarers more directly.

3. See in particular the articles by Asif A. Siddiqi, Slava Gerovitch, and Andrew Jenks in *Into the Cosmos: Space Exploration and Soviet Culture*, edited by James T. Andrews and Asif A. Siddiqi (Pittsburgh, PA: University of Pittsburgh Press, 2011), 47–132.

4. See Andrews and Siddiqi, *Into the Cosmos*, and *Soviet Space Culture: Cosmic Enthusiasm in Socialist Societies*, edited by Eva Maurer, Julia Richers, Monica Rüthers, and Carmen Scheide (London: Palgrave Macmillan, 2011). See also Cathleen S. Lewis, "From the Cradle to the Grave: Cosmonaut

Nostalgia in Soviet and Post-Soviet Film," in *Remembering the Space Age*, edited by Steven J. Dick (Washington, DC: NASA, 2008), 253–70.

5. Roger D. Launius, "Heroes in a Vacuum: The Apollo Astronaut as Cultural Icon," *Florida Historical Quarterly* 87 (Fall 2008): 174–209, quotations on 193–94. This article is the basis for a chapter in the forthcoming *After Apollo: The Legacy of the American Moon Landings* (New York: Oxford University Press, forthcoming). More recently, Matthew H. Hersch, a contributor to this volume, has published "Return of the Lost Spaceman: America's Astronauts in Popular Culture, 1959–2006," *The Journal of Popular Culture* 44 (2011): 73–92, which is a hybrid, discussing real astronauts but mostly their fictional representations by Hollywood.

6. Tom Wolfe, *The Right Stuff* (New York: Farrar, Straus, and Giroux, 1979).

7. Although not written from the standpoint of cultural history (the author is a political scientist), Howard E. McCurdy's *Space and the American Imagination* (Washington, DC: Smithsonian Institution Press, 1997) has been an important stimulus to work both on the United States and beyond, notably Asif A. Siddiqi's *The Red Rocket's Glare: Spaceflight and the Soviet Imagination, 1857–1957* (Cambridge: Cambridge University Press, 2010) and Alexander C. T. Geppert's edited volume, *Imagining Outer Space: European Astroculture in the Twentieth Century* (New York: Palgrave Macmillan, 2012). Another important work, written by a literary scholar, is De Witt Douglas Kilgore's *Astrofuturism: Science, Race, and Visions of Utopia in Space* (Philadelphia: University of Pennsylvania Press, 2003). NASA conferences cosponsored by the National Air and Space Museum have also been valuable. See the proceedings *Societal Impact of Spaceflight*, edited by Steven J. Dick and Roger D. Launius (Washington, DC: NASA, 2007), and *Remembering the Space Age*, cited above, although the former primarily takes a socioeconomic, rather than cultural, approach. However, in all of these works, images of actual spacefarers have scarcely been discussed.

8. Matthew H. Hersch, *Inventing the American Astronaut* (New York: Palgrave Macmillan, 2012), a revision of his "Spacework: Labor and Culture in America's Astronaut Corps, 1959–1979" (PhD diss., University of Pennsylvania, 2010). See also his "Return of the Lost Spaceman," 73–92.

9. Andrew L. Jenks, *The Cosmonaut Who Couldn't Stop Smiling: The Life and Legend of Yuri Gagarin* (Dekalb: Northern Illinois University Press, 2012). Few biographies of U.S. astronauts merit consideration here, but James R. Hansen's thorough and readable *First Man: The Life of Neil Armstrong* (New York: Simon and Schuster, 2005) does discuss the often bizarre dimensions of Armstrong's image post-Apollo 11, such as the widespread belief in the Islamic world that he heard the call to prayer on the Moon and subsequently became a Muslim.

10. Stephen Harrigan, *Challenger Park* (New York: Knopf, 2006).

11. Geppert, *Imagining Outer Space*, 8–9.

CHAPTER 1

Setting the Scene for Human Spaceflights: *Men into Space* and *The Man and the Challenge*

Margaret A. Weitekamp

As both the United States and the Soviet Union began working toward the first human spaceflights in 1961, two significant programs appeared on American television. On NBC, *The Man and the Challenge* (1959–60) used the inspiration of Air Force Colonel John Paul Stapp's well-publicized rocket sled experiments to create a series featuring a fictional doctor/researcher whose scientific experiments probed the limits of human endurance. On CBS, *Men into Space* (also 1959–60) depicted the realistic adventures of Colonel Edward McCauley, head of a fictional American space program. Aimed at adults and executed with the cooperation of the Department of Defense, *Men into Space* offered a fact-based depiction of space flight in the near future of the budding space age. Both programs were produced by Ziv Television Programs, Inc., a unique Midwestern company known as the leading producer of first-run syndicated programming.

Ziv's two television programs represented a transitional moment in American space-themed television, one that provides both context and background for understanding how the first human spaceflights were received. In contrast to the popular "space operas" of the early and mid-1950s, both series addressed the challenges of going into space in realistic ways. In particular, *Men into Space* depicted one military-inspired vision of how a spacefaring future—and its space travelers—could look. For its part, even without leaving the ground, *The Man and the Challenge* explored the question of what kind of person would be best qualified to take the first steps into space. In the end, both shows glorified the ingenuity of man's technological prowess as they pushed back boundaries. Because of their depictions of gender, both male and female, and by depicting adventures based on actual scientific principles and practices, they also prepared television audiences for the public performances of technological achievement that human spaceflight provided and the risks they entailed.

The transitional moment these programs represented was fleeting. Because of changes in how programming was being created, independently produced network television soon vanished. As a result, both *The Man and the Challenge* and *Men into Space* each ran for only one season, 1959–60, yet in that season both shows tapped into contemporary interest in the science fact about what was possible for the men (and perhaps women) who pushed the boundaries. Analyzing the depictions of gender, science, and technology in these two TV programs reveals one part of the cultural contexts that helped shape the reception in the United States of Yuri Gagarin's flight and, even more so, Alan Shepard's mission as the first American in space, both in 1961.

Early 1950s Space Television

To appreciate how and why Ziv Television developed the two series, it is important to understand the wide variety of program formats, production processes, and business models that coexisted in the early years of television. Space adventure programs from the early 1950s were often fantastical, cheaply produced, and melodramatic because in many ways the business models that eventually created high-quality episodic television were still developing. However, such programs from the early and mid-1950s provided the backdrop against which *The Man and the Challenge* and *Men into Space* would have initially been received.

The very first such program, *Captain Video and His Video Rangers* (1949–55), known as "TV's first space opera," debuted on the DuMont Network on June 27, 1949. An inexpensively produced series, the live show was broadcast five times a week for six seasons. The lead character of Captain Video, assisted by a teen-aged Video Ranger, carried out an ongoing battle against the evil scientist Dr. Pauli. Wearing a military-style jacket with a lightning bolt across the chest, Captain Video operated from a mountaintop hideout full of dials and switches. The character also had amazing devices, cobbled together on the program's paltry $25-a-week prop budget. Each episode also featured an incongruous interlude in which Captain Video would invite his viewing video rangers to watch clips of Western action pictures (drawn from the DuMont holdings). Such breaks both extended the show's length and allowed for set changes during the live program.[1]

DuMont's business model for *Captain Video* remained rooted in radio production, with more money invested in advertising than in production. Mail-away premiums from the show's sponsors were the show's first merchandise. As a manufacturer of TV sets, DuMont aimed ultimately to sell receivers, not to make money through the programs themselves. DuMont executives actually saw so little value in the recordings of *Captain Video* that they destroyed them to avoid storage costs, choosing instead to reclaim the silver content in the films. In the early years of TV programming, few people imagined that audiences might be willing to watch a television program more than once.[2]

ABC's long-running *Space Patrol* (1950–1955) reinforced the reputation of 1950s space programs as melodramatic. *Space Patrol* was a space-based police procedural, broadcast live. Wearing uniform tunics and peaked military-style hats, the *Space Patrol* cast had to memorize new scripts for each fifteen-minute daily show, later adding a thirty-minute live broadcast every Saturday. Live television (which accounted for 80% of network shows in 1953), although dynamic, had inherent limitations. For example, the space-themed special effects had to be done in real time. They could not be added or enhanced later. As a result, even with a significant production budget and talented staff, the effects could appear campy. Moreover, despite live television's reputation for conveying the excitement of live theater (and having a higher-quality broadcast picture than programs recorded using kinescopes, which filmed the picture shown on a monitor), such programming also held the potential for in-the-moment disasters. In one episode of *Space Patrol*, an actor playing a villain simply froze. The other actors improvised, pretending the character was now telepathic. Such scenarios were inherent in live TV.[3]

During this same time period, well-credentialed spaceflight enthusiasts (including such notables as author Willy Ley, engineer Wernher von Braun, and artist Chesley Bonestell) tried to compete with fantastical television programs by depicting spaceflight realistically. In *Space and the American Imagination*, Howard McCurdy argued that these men and several others, whom McCurdy called "space boosters," self-consciously used popular media to present human spaceflight as realistic and achievable, in marked contrast to the period's fantastical science fiction. Their varied campaign included work on George Pal's Academy Award–winning *Destination Moon* (1950), a famous series of articles in *Collier's* magazine from 1952 to 1954, and collaborations with Walt Disney to produce space-themed, nonfiction television programs in the mid-1950s as well as the iconic Rocket to the Moon ride at the center of Disneyland's Tomorrowland in 1955. The space boosters' efforts shaped the discourse about human spaceflight for decades to follow.[4]

Some of those influences found their way into regular episodic television programs. *Tom Corbett, Space Cadet* (1950–55) earned a reputation as "easily the most scientifically accurate" of the contemporary space shows. Ley served as a consultant, successfully lending his critical eye to script reviews. Although the cast occasionally donned spacesuits, the restricted effects budget kept them in space uniforms most of the time, wearing exaggerated flight suits with dotted necklines and deep Vs on the chests, accented with an emblem bearing a rocket ship and a lightning bolt. Despite five successful seasons, *Tom Corbett* had a rocky broadcast history. The program bounced from network to network, showing in successive seasons on four different television networks. CBS later created a copycat show, the short-lived *Rod Brown of the Rocket Rangers* (1953–54) starring a young Cliff Robertson, both to compete with DuMont's *Captain Video* and ABC's *Space Patrol* and to make up for allowing *Tom Corbett, Space Cadet* to go to ABC.[5]

Throughout the early and mid-1950s, the ongoing popularity of space-themed programs also inspired the resurrection of established space heroes. *Buck Rogers* and *Flash Gordon*, well-established multiformat franchises that had flourished in the 1930s, found new life in the early 1950s. ABC broadcast a live TV program of *Buck Rogers* (1950–51), for which no kinescope recordings survive. In 1954, a syndicated version of *Flash Gordon* (1954–55) was filmed in West Germany and France for syndication. By the mid-1950s, most television networks included a space show in their lineup.[6]

Even as late as 1955, however, the best formula for television production had still not yet been established. NBC's *Commando Cody, Sky Marshall of the Universe* (1955) was released first as a movie serial, demonstrating that television was still not viewed as an exclusive medium. To minimize costs, the show reused the Rocket Man costume from the Republic Film movie serial *King of the Rocket Men* (1949). Whenever the lead character (played by Judd Holdren) was not wearing the Rocket Man leather flight costume with its cumbersome metallic helmet, he wore a black military jacket with a rocket on the sleeve and a *Lone Ranger*–style domino face mask. Keeping Holdren's face partially covered presumably prevented him from gaining a following and demanding more money (and would have eased the transition to another actor if he left the show).[7]

The sets and costumes used in early 1950s space-themed TV programs reinforced the general impression that these were fantastic stories aimed at children, without any real scientific content. Most of the lead characters wore military-style uniforms accessorized with lightning bolts, rocket ships, or other symbols that suggested something futuristic. Likewise, the weapons and instruments used had elaborate technobabble names but little real explanation. Notably, most shows did not feature spacesuits or helmets. When space helmets were shown, as in the case of *Tom Corbett, Space Cadet*, their large bubble construction made it easy to see the actors' faces.

Examining the space operas of the early 1950s establishes not only the context against which later, more realistic, adult space programs would have been viewed but also the business environment in which they were produced. *The Man and the Challenge* and *Men into Space* were products of a particular set of circumstances in the structure of the television industry. Both shows were created by Ziv Television Productions, a company driven by Frederick W. Ziv, who found a niche in television and radio production, became expert in filling it, and adapted his business models as the industry changed. To understand how these programs originated and why they did not survive, one must understand Ziv, both the man and the companies.

Ziv

Before Frederick Ziv became known as the "father of broadcast syndication" and his companies became the "first financially successful producers of large-scale syndicated programming," he began as a young advertising executive in Cincinnati, Ohio. Looking to make his mark with his own agency, Ziv began targeting radio because, as a new medium, it presented a better opportunity for a newcomer than magazine or newspaper advertising did, being already dominated by established firms. Therefore, Ziv first developed what would become his lifetime business model by working in radio.[8]

Rather than simply buying advertising on established programs or stations, Ziv convinced his local clients that they could make greater impressions by sponsoring independently produced programs. Ziv's radio shows, pitched by his agency's salesmen to various local sponsors, included not only the programs themselves but also detailed plans for all of the promotional materials required to garner enthusiasm for the program, allowing them to compete against better-funded network shows. By 1948, even as his company continued to work in radio (producing more than twenty different radio series for national syndication), Ziv expanded this approach to include television.[9]

Like radio before it, TV relied on networks. Belonging to a network brought not only built-in programming but also prestige to a station. Network programs with high production values funded by national advertisers could only be broadcast by the group of stations affiliated with that network. (The networks favored live programming at least in part because it meant individual stations had to be affiliates to receive the one-time broadcasts, putting independent stations at a relative disadvantage.) But stations also needed programming during the non-prime-time hours for which the major networks did not provide it, and independent stations needed programming as well. Ziv Television Productions, Inc., which always remained based in the Midwest—eventually opening offices in New York and Hollywood but keeping its headquarters in Cincinnati—specialized in creating independent shows to be "syndicated" or sold directly to individual stations. Ziv eventually became known as the "most prolific producer of programming for the first-run syndication market during the 1950s." Aided by a law degree from the University of Michigan (Ziv envisioned every sales pitch as a legal argument, anticipating counterarguments and preparing rejoinders beforehand), he was a savvy businessman who found his niche in syndication.[10]

A forward-looking entrepreneur, Ziv recognized emerging trends in television production. For instance, Ziv recorded the company's first major television series, *The Cisco Kid* (1950–56), a Western, on color film, even though neither recorded programs nor color sets were in wide use when the series began (nor, for that matter, even when it ended). But because the program was filmed in color and because Ziv foresaw a market for rebroadcasting television programs, *The Cisco Kid* survived in reruns well into the 1970s, outliving even Ziv's company.[11]

Ziv's plan for success in television syndication focused on reliable, entertaining television programs in proven genres. As Frederick Ziv recalled in 1975,

> It was obvious to all of us who had our fingers on the pulse of the American public that they wanted escapist entertainment . . . We [Ziv] did not do highbrow material. We did material that would appeal to the broadest segment of the public. And they became the biggest purchasers of television sets.[12]

Ziv sought out stars who needed no introduction, such as comedians Red Skelton and Eddie Cantor for variety programs, and proven formulas for dramatic pieces. As a result, Ziv Television specialized in half-hour male-oriented action adventure dramas, including such reliable genres as Westerns and crime or courtroom dramas, each written with a slight twist to keep them interesting. By the mid-1950s, Ziv was increasingly convinced of the TV audience's affinity for realistic programs, a trend that was reinforced in his mind by the financial success of *Science Fiction Theater*.

Produced for Ziv by Ivan Tors, who would later produce *The Man and the Challenge*, *Science Fiction Theater* (1955–57) was hosted by Truman Bradley, a former radio announcer. Each episode began with a science experiment, the principles of which inspired the drama that followed. Ziv believed that the success of this adult series would depend upon the veracity of the science (and thus the willingness of technically savvy industry sponsors, such as Conoco in Dallas, Texas, to support the show). As a result, the production team included a six-person research department with a $75,000 budget for fact checking. The formula worked. For each of its 78 episodes, *Science Fiction Theater* was always first or second in the ratings. Such success reinforced for Frederick Ziv that "science fact" shows aimed at adults could be a bankable approach.[13]

The turn to more adult programming was in line with what other contemporary television producers were doing. For instance, in the late 1950s, Westerns aimed at adults, such as *Gunsmoke* (1955–75) and *Bonanza* (1959–73), became all the rage. Six such shows appeared in 1955–56, followed by eighteen the next year. Their successes created a boom. In the 1959–60 season, thirty different programs set in Western locales appeared on the air. The Western frontier motifs that were so popular on television even informed how policymakers framed the U.S. space program.[14]

Ziv's productions for 1959–60 aimed not only to capitalize on the trend toward adult genre programming but also to capture the contemporary interest in space exploration sparked by the space age (begun by the first artificial satellite, Sputnik 1, on October 4, 1957). Having a recipe for successful programming became increasingly important to Ziv as his company worked to compete in a rapidly changing television market. By 1956, TV had established itself as a successful medium: 72% of all American households had a set. But network executives had also learned how to make money with the programs themselves, rather than relying on just advertising or set sales. In 1956, the major networks began syndicating their own programs. To keep pace, Ziv Television changed its tactics. Rather than marketing independently produced programs to individual stations, Ziv began producing programs directly for the networks. Although he personally resented them, preferring independent syndication, the company's producers and writers were flattered to have their work broadcast on a major network, and he recognized the business value of the new direction.[15]

Men into Space

Men into Space was one of those offerings, produced directly for CBS and first appearing at 8:30 p.m. on Wednesday nights in the fall of 1959. Rather than a fantastical space opera, *Men into Space* was in many ways an extension of the space realism seen in programs such as *Tom Corbett, Space Cadet* and the work of space boosters from earlier in the decade. With high-caliber technical advice and plot supervision, *Men into Space* aimed "to give the public accurate, nonclassified information on accomplishments, operations, known and anticipated hazards in space" (Figure 1).[16]

The Department of Defense agreed to support *Men into Space* as long as script approval was included. Captain M. C. Spaulding from the U.S. Air Force Ballistic Missile Division served as a technical advisor. The famous space artist Chesley Bonestell, who had published space paintings in *Collier's* and *Life* in addition to doing extensive work for Hollywood movies, developed the "space concepts," or production design, for the program. When it was broadcast, the final credits for *Men into Space* acknowledged the Defense Department, as well as the Air Force's Air Research and Development Command, Office of the Surgeon General, and School of Aviation Medicine. The reliance on the Air Force reflected contemporary expectations (and Air Force hopes) that the service would be the natural military branch to control future human spaceflight.[17]

Ziv recognized the value that subject matter authorities brought to a program. Beginning in 1953, the promotion for the anticommunist triple-agent

FIGURE 1. With authenticity bolstered by the Department of Defense's script reviews, Ziv Television's *Men into Space* (1959–60) depicted a robust human spaceflight program that included regular trips to the Moon by crews of multiple astronauts, such as this scene from "Water Tank Rescue" (October 28, 1959). *Men into Space* © Metro-Goldwyn-Mayer Studios, Inc. All Rights Reserved. Courtesy of MGM Media Licensing.

drama *I Led Three Lives* (1953–56) emphasized that the FBI reviewed each script, which were based on the writings of FBI counterspy Herbert Philbric. Ziv also cooperated with the U.S. Military Academy to produce a true-to-life show called *West Point* (1956–58). Likewise, Ziv's successful *Sea Hunt* (1958–61) received technical assistance from the Scripps Institute of Oceanography, University of California. The participation of recognized space experts reinforced the premise of *Men into Space* as a fact-based depiction of the near future in spaceflight.[18]

Originally slated to be called *Moon Probe*, *Men into Space* featured a rotating cast of various support crew (astronauts and other scientists) making repeated missions to the Moon, led by Colonel Edward McCauley (portrayed by Bill Lundigan). After an opening montage of stock footage of a V-2 (the captured German liquid-fuel rocket) and glimpses of the fictionalized characters who would be the program's heroes, a voiceover explained the program:

> The story you are about to see has not happened, yet. These are the scenes from that story. A story that *will* happen as soon as these men

are ready. This is a countdown. A missile is about to be launched. It will be the XMP-13. "XMP" meaning eXperimental Moon Probe. A missile that will carry three human beings into outer space.

Variations of the dialogue about a story that would soon be true, once the men were ready, introduced each episode. In the opening lines of dialogue of the first episode, McCauley gave his eagle colonel's insignia to his son Pete (portrayed by Charles Herbert) to hold for safekeeping until he returned from the mission. The music swelled as McCauley kissed his wife, Mary (played by Angie Dickinson), goodbye. McCauley's interactions with his family throughout the series continued to reinforce the humanity of the men going into space.[19]

Having a press conference depicted within the first episode allowed the mission to be explained to the viewing audience, both before the launch and during the flight. Next, the slow preparations for the vehicle's launch (including, in the first episode, counting down steadily from forty) featured continual cuts between the men in the vehicle and the men at mission control. Stock film of various rocket launches provided the footage of the fictional launch. But the episode's drama finally developed when the rocket's second stage failed to separate. McCauley had to go outside the vehicle to cut it loose as the military leaders on the ground continued to explain the scientific principles to the assembled audience (and thus to the viewers). Ultimately, McCauley was knocked away from the vehicle but was rescued after tracking stations around the world, including a Soviet station, helped to locate him for retrieval. In the end, international cooperation became one of the episode's final lessons. As McCauley reminded another officer during the mission debriefing, "For one half hour, the entire world made one human life more important than anything else."[20]

In addition to learning something about realistic plans for spaceflights and the risks that could be faced, viewers watching these episodes heard rationales for going into space. Within the first few minutes of the first episode of *Men into Space*, a newspaperman asked McCauley straight out, "Why?" McCauley's reply evoked British mountaineer George Mallory's famously simple answer about climbing Mount Everest: "Because it's there." McCauley addressed the press conference with, "If a mountain [exists], somebody has to climb it. The mountains on the Moon just happen to be a few hundred thousand miles higher. Let's call it a way of life . . . Let's say science is a way of life." McCauley's answer provided the first of many justifications for human spaceflight *Men into Space* would provide.[21]

In the second episode, "Moon Landing," two additional reasons for making a space voyage arose in the first few minutes. After a briefing, a senator, introduced as the chairman of the president's space committee, questioned the cost: "Do you know how many billions are being spent to get you four men up there?" "Yes, sir, I do," McCauley replied, "And I believe it is a good investment. As you know, practical applications to Moon conquest are enormous." Although McCauley did not offer any details to buttress his claims, his colleague quickly offered a more philosophical rationale for the flight. Dr. Russell answered,

> Well, nothing stands still, Senator. Life began in the sea, groped onto the land, and with intelligence and time, it staggered into the sky. Now we are leaping into space. We are ready. Spaceflight is only a natural, inevitable step in evolution. We have to go sooner or later. We might as well make it sooner.

To reinforce the point, McCauley concurred, "He's right."[22]

The rest of the second episode underscored the value that McCauley and his crew placed on the mission, as well as an additional practical application of the trip. While exploring the lunar surface, McCauley explained how the lunar atmosphere's clarity would facilitate the astronomical observations made from a Moon-based telescope. But deploying the telescope had a high cost. While setting it up, Dr. Russell, the character who spoke so eloquently about the need for spaceflight in the episode's opening minutes, collapsed from injuries sustained during the trip, the severity of which he hid from McCauley because he wanted to complete the mission. After his death, as his last request, the crew left Russell on the lunar surface next to the telescope. Later, back on Earth, Russell's mother tells McCauley that she is satisfied that her son died doing what he wanted to do. Russell's death illustrated the officer's dedication to the spaceflight cause and the integral roles that scientists would play in a future military astronaut corps. (*Men into Space* did not anticipate the distinctions between scientist astronauts and pilot astronauts that became so important in the actual civilian astronaut corps, emphasizing instead the unifying military affiliations of its crew members.)[23]

The question of putting lives at risk emerged again in an episode titled "Water Tank Rescue." An opening scene, set with an astronaut's family at his home, dramatized the risk to the men involved by contrasting the son's naïve enthusiasm (comparing the upcoming mission to a "milk run") with the wife's worried face. Her concerns were borne out when her husband had a heart attack on the Moon. Doctors on Earth quickly determined that the force of launching home would kill him. Fortunately, McCauley and the rest of the crew were able to improvise a water tank device that would supposedly cut the acceleration forces and bring him home safely. Although the science behind this solution was ultimately faulty, the show illustrated how ingenuity could compensate for distance, reinforcing the broader point that the United States has the capabilities to solve potential problems.

FIGURE 2. Lead actor William "Bill" Lundigan posed for this *Men into Space* publicity shot wearing the show's costume space suit, modeled on real examples of contemporary military flight suits and prototype spacesuits. *Men into Space* © Metro-Goldwyn-Mayer Studios, Inc. All Rights Reserved. Courtesy of MGM Media Licensing.

Although initially dressed in real Navy full-pressure suits (the credits for the first episode noted "space suits worn in outer space sequences [were] provided by the United States Navy"), the show's cast quickly switched to fictional costumes based on state-of-the-art spacesuit designs (Figure 2).[24] The head-to-toe silver costume spacesuits worn by McCauley and his crew were almost certainly inspired by actual prototype spacesuits that had been publicized as early as 1957. The silver X-15 suit was a full-pressure high-altitude flight suit developed by David Clark for pilots to wear when flying the X-15, a joint Air Force–National Aeronautics and Space Administration (NASA) experimental rocket plane that could reach the edge of space. Unveiled to the public in a *Los Angeles Times* article in 1957 and featured on the cover of *Life* magazine in January 1958, the X-15 suit featured a coverall of aluminized nylon (nylon with a vacuum-blasted aluminum coating). Whether that layer actually provided additional heat resistance or radiation deflection (as the suit's designers claimed) or if, as X-15 test pilot Scott Crossfield later recalled, it simply made the suits look more photogenic, "like a spacesuit should," the silver-colored spacesuit became the image of the future space traveler. The famous Project Mercury spacesuit developed in 1959 by Goodrich for NASA also featured an aluminized nylon outer layer.[25]

The images of the spacefarer depicted by McCauley and his space crew suggested both high-tech realism based in contemporary spaceflight research and the potential interchangeability of spaceflight participants. In each episode of *Men into Space*, McCauley remained the central character, but other crewmates proved to be largely interchangeable. New characters and new actors appeared each week. Dressed in identical silver spacesuits with full helmets that carried an Air Force logo, the supporting characters became hard to distinguish from each other, especially during their fictional Moon walks. To ease this complication, each of the *Men into Space* astronauts had his surname stenciled on his space helmet above the clear visor so that viewers could tell them apart.

(Facing an analogous problem, NASA later added a red stripe to the spacesuits worn by Apollo mission commanders in order to distinguish them from their fellow moonwalkers in photo and video records.)

As the series progressed, the plots for *Men into Space* dramatized emergency scenarios that the Department of Defense reviewers determined to be believable. The rescue of a rogue spaceship, the defusing of a nuclear power plant aboard an unmanned missile, a collision between a spaceship and a refueling tanker, and a runaway satellite all put McCauley and his crewmen to the test (it was an adventure series, after all). Week after week, viewers saw McCauley and his various compatriots persevering despite significant risks. Whether endangered by their own actions or from unforeseen phenomena, McCauley and the fictionalized American space program depicted in *Men into Space* pursued spaceflight's benefits, which included an orbiting space station and regular crews stationed on the Moon. The message that the benefits outweighed the risks was clear.

The Man and the Challenge

Three weeks before *Men into Space* appeared on TV, *The Man and the Challenge* began the broadcast of what would eventually be thirty-six black-and-white half-hour episodes. Produced by Ivan Tors for Ziv, the show aired on NBC on Saturday nights at 8:30 p.m. The show's premise was inspired by aeromedical researcher John Paul Stapp and his contemporaries, who personally tested how the human body would react to the predicted rigors of spaceflight (Figure 3). Originally, the program was going to be titled simply *Challenge*. But Ziv's producers worried that the show would be confused with a game show, evoking the quiz show scandals of 1958 and 1959. As a result, they added "The Man" to the title, a change that ultimately benefited the program by putting the emphasis on the central character, Dr. Glenn Barton (played by George Nader) of the fictional Human Factors Institute.[26]

The opening titles reinforced the character's parallels with the adventurous Stapp, who famously strapped himself into a rocket-propelled sled with water brakes in order to test the human body's reactions to acceleration and deceleration. Each episode began with a darkened silhouette of a man sitting in an instrumented chair on a rocket sled. As the first part of the title, "The Man," flashed on screen, the profile illuminated to reveal the main character, Nader as Barton, donning a crash helmet with a darkened visor. As "and the" appeared, a film clip of Stapp's well-publicized rocket sled experiments began. The last part of the show's title, "Challenge," appeared superimposed over plumes of spray from the water reservoir. The opening title ended with a close-up of Nader removing a water-spattered helmet to reveal his face (which, unlike Stapp's, remained unbruised), a final focus on the man at the center of

FIGURE 3. *The Man and the Challenge* drew inspiration from contemporary aerospace researchers such as Air Force Colonel John Paul Stapp. Stapp's personal participation in rocket sled experiments, pictured here, inspired the program's title sequence. Courtesy of AFTC History Office.

the show. Each week, the episode that followed tackled a dramatized problem based on contemporary research on human physiology.[27]

Although *The Man and the Challenge* never had an episode with an outer space setting, the question of imminent human spaceflight loomed over much of the series. In an episode titled "Experiments in Terror," when Glenn Barton found himself the unwitting subject of several tests of his reactions, nerve, and bravery, the doctor overseeing the tests explained directly, "We need . . . highly-specialized men, who will one day land on the Moon and neighboring planets. Who can withstand pain, terror, cold, heat, hunger, sleeplessness, weightlessness, and isolation."[28] Likewise, when Barton subjected two elite pilots to the unusual endurance test of entering them in a daylong cross-country stock car race, he explained, "They were willing to try to undergo any experiments deemed important for space progress."[29] Like the researchers on which the show's premise was based, the fictional Barton conducted his experiments in spaceflight readiness using elaborate simulations.

Different iterations of episode scripts, now housed in the United Artists collection at the Wisconsin Historical Society in Madison, reveal that an even greater number of spaceflight references appeared in early versions of scripts. Used to justify the action in initial drafts, explicit references to spaceflight sometimes disappeared in the final filming or editing. This was especially true once the program had a following and the show's premise had been established. In later episodes, the action sequences required less supporting explanation.[30]

Because the series' premise remained earthbound, the only spacesuits appeared in an episode titled "The Visitors." In that installment, three experimenters, including Barton, traveled to a remote desert testing site to field test prototype "planet suits." Unlike the spacesuits shown in *Men into Space*, however, these imagined prototype suits featured large helmets with elaborate antennae. In the plot's dramatic twist, poachers trespassing on government land concluded that the researchers were space aliens and hunted them with rifles. Because the hermetically sealed suits supposedly required the wearers to be in a pressure chamber to remove them safely, the men had to rely on their wits until their rescue. Despite the lack of spacesuits, throughout its run *The Man and the Challenge* directly explored the question of what kind of person should be an astronaut.[31]

In fact, the very first episode of the series explored the qualities needed for spaceflight in both men and women. In "Sphere of No Return," Dr. Glenn Barton appeared inside a spacecraft simulator in a pressure chamber with two male test subjects. As three men worked through exercises, Barton explained the experiment in a voiceover: "One danger man will have to face in space is a sudden puncture of his spaceship by a meteorite." Barton then raised a gun (behind and out of sight of the two male subjects) and shot the thick glass porthole (Figure 4). Immediately, the simulated craft, presumably inside a vacuum chamber, began to vent atmosphere. As Barton observed, one of the men tried to plug the hole with a large suction device but fumbled it. Barton's voiceover dramatically identified the problem, which was not the device but the man: "Mason panicked." Demonstrating the cool competence that defined the series' lead character, Barton completed the repair. Mason, who had otherwise passed all preliminary tests with flying colors, had to be replaced. Spaceflight required both physical and psychological toughness.[32]

The focus on sorting out psychological weaknesses contrasted with the striking visual display of physical masculinity. For reasons that were never explained—and never seemed to be necessary except to emphasize their bodies—all of the men in the spacecraft simulator, including Barton, trained with their shirts off, bare chested. Despite the sensors taped to their chests, their headband receptors were apparently tracking their vital signs, so the bare chestedness was not plot related. Nonetheless, each of the physically strong young men wore only tight shorts and athletic shoes and socks. To the viewer,

FIGURE 4. In this scene from *The Man and the Challenge*'s first episode, George Nader as Dr. Barton (left) prepares to shoot a hole in a spacecraft simulator to test the mettle of two subjects. Spaceflight required physical and mental toughness, traits often equated with rugged masculinity at the time, a relationship that this episode questioned. *Man and the Challenge* © Metro-Goldwyn-Mayer Studios, Inc. All Rights Reserved. Courtesy of MGM Media Licensing.

they all appeared to be equal examples of muscular manhood until the experiment revealed Mason's weakness.

But *The Man and the Challenge* also investigated whether women might possess the qualities needed for spaceflight. As the character of Glenn Barton explained in "Sphere of No Return," "if there are going to be colonies in outer space, the pioneers can't all be men." Appropriate female test subjects were easily found at Barton's Human Factors Institute. The episode's opening shot showed three men in lab coats sitting at consoles, while Barton's assistant Miss Allen (played by Joyce Meadows) occupied another console in the foreground. Rather than test her openly, Barton initially measured Allen's abilities by taking her on a date to an amusement park, where he surreptitiously timed her reactions using a stopwatch.[33]

When Barton decided to include Allen in spaceflight readiness experiments, the romantic overtones of Miss Allen's obvious crush on Dr. Barton overshadowed her participation—and established him a charismatic and romantically attractive leading man. Having measured her abilities at the amusement park, Barton included Allen in a series of tests, pairing her with

Cory, a male subject who had also already proven himself. Throughout a pressure chamber simulation test and a real high-altitude balloon flight, both Allen and Cory experienced the psychological dangers of spaceflights. The group's success through a series of adventures in the gondola (avoiding power lines, dodging thunderstorms, and surviving hypoxia) ended with a safe landing thanks to Dr. Barton's quick thinking. In the final scene, Miss Allen confessed her attraction. Both she and Barton realized, however, that his dedication to work made a relationship impossible.[34]

The question of women's physical capabilities recurred several times throughout the series. In an episode titled "Escape to Nepal," Barton included an expert laboratory technician and linguist named Marilyn Sidney (portrayed by Joan Granville) in his high-altitude mountain-climbing expedition because "one day women might be indispensable in spaceflight." Although Sidney also made sandwiches and coffee as the group climbed, she demonstrated her value to the expedition through her linguistic skills.[35] Barton again tested women for possible spaceflight stresses in an episode titled "Astro Female." After narrowing down his possible candidates from the hundreds who volunteered, Barton tested four and concluded, in a voiceover describing the reaction of another colleague, "Dr. Cremer's eyebrows went up. He couldn't believe it. But it was proven. What I had always believed: the female body is even better equipped to handle radical changes than the male's [is]." The show concluded with the summary statement that "when the chips are down, the so-called 'weaker sex' is a myth."[36] Such episodes reflected contemporary scientific curiosity about women's physical abilities even as American society reinforced narrow gender roles, especially for married, middle-class white women.[37]

In contrast, women in *Men into Space* served primarily to express fears unacknowledged by men. When a lead researcher's wife questioned the safety of flying a nuclear-powered rocket in "Lost Missile," her husband dismissed her question before it could be answered, saying to McCauley, "Now, isn't that just like a woman?"[38] Such fears had no place in a spacecraft. In an early episode, "Moon Landing," McCauley offered the astronauts' wives the backhanded compliment that as a man, he would not be able to endure such worrying; he would much rather fly to the Moon than sit at home, watching the risky venture. When they were not reflecting the emotion of the drama, however, women in *Men into Space* represented dangerous distractions. In an episode titled "Moonquake," an astronaut preoccupied by concerns about his injured wife, who had been in an automobile accident, put the crew in danger.[39] Likewise, in "First Woman on the Moon," the only episode of *Men into Space* with a woman shown going into space, Renza Hale, the wife accompanying her scientist husband on a mission, caused problems for the crew when she refused to acknowledge the need for her own protection.[40]

In *The Man and the Challenge*, the exploration of gender and capability became metaphorical as well as literal. In the second episode, "Maximum Capacity," Dr. Barton worked with three world-class American male skiers to answer the question, "Are skiers better qualified to function and survive among certain extreme conditions?" Quickly, however, the episode revealed itself to be concerned with spaceflight—and national spirit. When one skier was afraid of a slope called Madden's Ridge, described as "a steep three thousand foot drop," Barton exhorted him to act by appealing to a national need for psychological toughness:

> Let's change the name of Madden Ridge. Let's call it outer space. We gonna give that up too without a try? We weren't the first with jets. We weren't the first with a satellite. Somebody else was. The North Pole is our new frontier. Our radar posts are scattered all over the Arctic. Finland was saved once by the quality of her ski troops. Could we do the same thing if it happened in Alaska? Or are we just going to keep on being second best from here on in?[41]

The message was clear: getting the skier to attempt an extreme slope modeled the national psychological toughening required for cold war success. According to these episodes of *The Man and the Challenge*, spaceflight would require a particular kind of toughness, both physical and metaphorical, often coded as masculinity and tied to national prestige.

Different Models of Masculinity

Such explorations reflected the cultural context. The late 1950s and early 1960s were a historical moment of gender anxiety. The rigid gender roles asserted after the end of World War II, described so effectively by historian Elaine Tyler May in *Homeward Bound*, seemed to be breaking down. What was called protofeminism, the first evidence of what would become the Second Wave of the women's movement, was being felt. At the same time, men suffered from a crisis in masculinity created by other cultural aspects of 1950s suburban life, including "momism," "organization men," and bureaucratic softness. The reactions to the "Sputnik moment" that called for a return to American frontier ideals also underscored cultural concerns about a loss of masculine toughness. Historian Robert D. Dean has argued that President John F. Kennedy's foreign policy cannot be understood without appreciating how ideologies about masculinity, acted out through cold war assessments of strength and weakness, underlaid his administration's decision making.[42]

A reassertion of traditional masculinity became a central message of several episodes of *The Man and the Challenge*. In "Odds Against Survival," Barton brought three prominent scientists and their wives aboard a nuclear submarine under the pretense that he had been ordered to rescue them from nuclear annihilation prompted by a European conflict. The scientists slowly adjusted to their confinement and began to seek survival solutions (which was, of course, the real test: how people could survive the extended confinement thought to be needed to outlast the aftereffects of a nuclear blast). At the same time, one of the scientists, the henpecked Dr. Robinson, learned to assert himself. Only as he took on a more traditional manly role was he able to help save the group—and his marriage.[43]

However, headstrong, unyielding cockiness was explicitly identified in several episodes as undesirable for space flights. In "White Out," Barton explained in the opening voiceover that "with manned interplanetary flight only years away, the problem of personnel for spaceships was becoming more acute. What was the best personality, physical, and psychological, for men in command positions?" Through a high-seas experiment with several highly capable test subjects aboard a sailing ship, Barton exposed one of them, John Napier (played by Keith Larsen), as overly assertive. Napier assumed too much responsibility, relied too much on his own supposed superiority, and ultimately proved unable to cope with even small setbacks, a dangerous combination for leadership in space.[44] Likewise, in an episode titled "Hurricane Mesa," Barton required a test pilot to become a clerk for ejection seat tests using dummies on a rocket sled. The man's successful adaptation to the boring job demonstrated that careful record keeping would be just as vital as human risk taking in the new space age.[45] Finally, an episode about the psychological conditions astronauts could face while confined in a floating spacecraft awaiting an ocean rescue ended with the reassertion of the importance of psychological screening for all space candidates.[46] The strength and tenacity needed for spaceflight must be a certain kind of masculinity, not uncompromising machismo.

In *Men into Space*, Colonel Ed McCauley exemplified contemporary masculine ideals (Figure 5). In addition

FIGURE 5. Bill Lundigan portrayed Colonel Ed McCauley as an ideal 1950s man: level-headed, action-oriented, and unfailingly loyal to family and country. *Man and the Challenge* © Metro-Goldwyn-Mayer Studios, Inc. All Rights Reserved. Courtesy of MGM Media Licensing.

to being a calm, deliberate leader of men (a model military officer), he was also a faithful husband and family man who bought his son a model rocket because he had to miss one of the boy's little league baseball games because of a mission. One scholarly analysis of 1950s television summarized the Colonel McCauley character as a reflection of the ideals of the time:

> McCauley represented one of those quiet, heroic figures which have now gone out of style. As the series developed, he evolved into almost a perfect paradigm of the way America was then pleased to view itself. At a time when the military was held in high esteem, McCauley was proudly exhibited as the best that the military had to offer. His presence was a reassurance to the nation that its welfare was in strong capable hands.[47]

In fact, Bill Lundigan, the actor who played McCauley, sometimes asked for script changes when he thought that the character was being depicted as too perfect to be believable. In the interest of dramatic tension, McCauley's crewmates had more license to be imperfect.

In comparison, Dr. Glenn Barton of *The Man and the Challenge* modeled an adventurous and challenging masculinity that was fit, capable, daring, and flirtatious—but ultimately married to his work. He was handsome and dashing. He was also somehow immune to the physical challenges that his subjects faced (even when he shared the same space with them). In the first episode, in which Dr. Barton tested people aboard an unpressurized balloon gondola, he watched, utterly unaffected by the thin atmosphere, as the other test subjects struggled to breathe, focus, or move. In most perilous situations encountered in the show, Barton provided the model for bravery, wits, and strength that allowed the situations to be resolved (Figure 6).

FIGURE 6. The dashing, dedicated, and daring scientist Dr. Glenn Barton was depicted by George Nader, who was in his personal life as openly gay as one could be in Hollywood in the 1950s. *Man and the Challenge* © Metro-Goldwyn-Mayer Studios, Inc. All Rights Reserved. Courtesy of MGM Media Licensing.

However, the muscular masculinity on weekly display in *The Man and the Challenge* remains open to multiple readings, especially in retrospect. For most of his adult life, George Nader was as openly gay as one could be in Hollywood in the 1950s. He met his life partner, Mark Miller, in 1947. They set up a household together, remaining together as a couple for 55 years, until Nader's death in 2002. Although neither came out publicly until 1985 after the death of their close friend and Nader's former Universal Studios colleague, Rock Hudson, Nader and Miller lived together openly. Nader resisted some of the studio's most extreme suggestions for disguising his sexual orientation, participating in public dates with Hollywood actresses but refusing to consider a sham marriage to a female secretary. Cast as a beefcake actor, Nader ultimately became frustrated with the opportunities Universal offered him and went out on his own in 1958 (thus making him available to star in the Ziv production of *The Man and the Challenge* in 1959). By 1964, he and Miller moved to Germany to explore Nader's options overseas. Nader's identity as a gay man would not have been known publically when *The Man and the Challenge* aired, but the assertive heterosexuality emphasized in his performance as Dr. Glenn Barton—and, indeed, the assumption of straightness at the core of most contemporary assertions of masculinity—is ironic in hindsight.[48]

Setting the Stage for Spaceflights

Projections of masculine strength, as demonstrated by McCauley in *Men into Space* and as examined by Barton in *The Man and the Challenge*, proved to be critical contexts for the first human spaceflight missions. The first such flight, Yuri Gagarin's single orbit on April 12, 1961, came just five days before the United States' very public failure in the invasion of Cuba by covertly trained Cuban exiles at the Bay of Pigs. The combination of two perceived defeats undermined President John F. Kennedy's image of taking a strong stance against worldwide communism.

Kennedy's cold war foreign policy was often cast, at the time and by the administration, in gendered terms of masculine strength. As Robert D. Dean has argued about Kennedy's diplomatic choices,

> Fear of the consequences of being judged "unmanly" influenced the reckoning of political costs or benefits associated with possible responses to those threats. In this sense, gender must be understood not as an independent *cause* of policy decisions, but as part of the very fabric of reasoning employed by officeholders.[49]

President Kennedy and his New Frontiersmen saw the space race in the same framework of strength versus weakness that imbued the rest of his foreign policy.

Alan Shepard's successful suborbital mission on May 5, 1961, became part of Kennedy's decision calculus about international displays of technological strength. Even before Shepard flew, White House decision makers were debating the nation's space policy. A week after Gagarin's flight, President Kennedy sent a memorandum to Vice President Lyndon B. Johnson asking, "Is there any other space program which promises dramatic results in which we could win?" Johnson's reply confirmed Kennedy's concerns about American prestige abroad and suggesting possible directions for the U.S. space program. In response, Kennedy declared before a joint session of Congress that the United States would complete a human lunar landing "before this decade is out." Amazingly, when the president committed the nation to that ambitious goal, American human spaceflight experience totaled fifteen minutes and twenty-eight seconds (the duration of Shepard's mission just three weeks earlier). The lunar landing decision illustrated how deeply Kennedy felt the imperative to exhibit strength, not weakness, in the cold war.[50]

Would ordinary Americans who watched these historic events unfold in 1961 have linked them to what they saw on television during a prime-time space-themed drama broadcast over a year earlier? Did they remember the depictions of masculinity in *Men into Space* or *The Man and the Challenge* as they heard about Gagarin's or Shepard's flights? Probably not directly. But the social and cultural contexts that guided Frederick Ziv as he produced those programs also shaped the conditions under which high-level political leaders considered American prestige and made real decisions about actual spaceflights.

In their short lifetimes, *The Man and the Challenge* and *Men into Space* reflected the contemporary state of thinking regarding spaceflight's risks and those who would soon assume them. Five months before these shows first aired, NASA had already announced that seven male jet test pilots would be the United States' first astronauts. But by the time Gagarin and Shepard flew their history-making missions, *Men into Space* had aired the rationales for spaceflights and justified the risks, and *The Man and the Challenge* had dramatized the physical and psychological demands that future space travelers would face. Together, these programs emphasized assertions of toughness and determination, coded as masculinity, both for individuals and for the nation.

The Coda, or How Does This Story End?

Men into Space and *The Man and the Challenge* represented a very different cultural moment than the space-themed shows that followed in the 1960s. By the time human spaceflight was well underway, viewers looking for realistic television programs about outer space found instead family sitcoms with space themes, such as *The Jetsons* (1962–63), *My Favorite Martian* (1963–66), *Lost in Space* (1965–68), and *I Dream of Jeannie* (1965–70). Such programs recast changing family and gender dynamics in space-themed settings but did not consider seriously the scientific underpinnings of spaceflight.[51] Although *Star Trek* (1966–69) did treat spaceflight as a central issue, the program still used its space setting primarily as a way to address contemporary social and cultural issues back on Earth.

Why was the serious consideration of human spaceflight on prime-time television so brief? For Ziv, the answer was that the industry changed. After the major television networks began producing their own syndicated programs, opportunities for independently produced series diminished quickly. In 1956, when the networks entered this market for the first time, there were twenty-nine first-run (independent) syndicated programs on the air. Four years later, there were ten. By 1964, only one remained. Ziv Television did not survive the transition.[52]

As Frederick Ziv recalled, by the end of the 1950s, the networks were demanding business agreements that made it impossible for independent production companies to compete, ultimately leading to his decision to sell.

> The reason I sold my business is because I recognized that the networks were taking command of everything, and were permitting independent producers no room at all . . . The networks demanded a percentage of your profits, they demanded script approval, cast approval . . . You were just doing whatever the networks asked you to do. And that was not my type of operation. I didn't care to become an employee of the networks.

Proud of his company's independence, Ziv chose to leave the industry rather than compete on disadvantageous terms. In July 1959 (just months before *Men into Space* and *The Man and the Challenge* went on air), he sold an 80% interest in Ziv Television to investors for a reported $14 million. A year later, United Artists acquired both the final 20% from Ziv and the initial 80% share from the banking groups. Frederick Ziv retired in 1965, after thirty-five years in the business.[53]

Would *Men into Space* or *The Man and the Challenge* have lasted more than one season if Ziv Television had survived to promote them? Perhaps. But *The Man and the Challenge* had stiff competition, airing each week opposite the iconic and, by that time, firmly established family situation comedy *Leave It to Beaver* (1957–63) on ABC. Also, *Men into Space* suffered from some of the earnestness and rigidity that was a by-product of official Department of Defense review. Although they left the airwaves after one season, thanks to Ziv's business model, both shows survived in reruns, giving them another chance to make an impression on their audiences. In the end, by depicting realistic adventures based on actual scientific principles and practices, *Men into Space* and *The Man and the Challenge* prepared television audiences for the technological achievements and public performances of risk-taking masculinity that human spaceflight soon provided.

Film Sources

The Library of Congress in Washington, D.C., holds six episodes of *The Man and the Challenge* in its Copyright Collection. A complete run of both programs are in the United Artists Corporation Records, which are held by the Wisconsin Center for Film and Theatre Research at the Wisconsin Historical Society in Madison.

Notes

1. Patrick Lucanio and Gary Colville, *American Space Science Fiction Television Series of the 1950s: Episode Guides and Casts and Credits for Twenty Shows* (Jefferson, NC: McFarland & Company, 1998), 96–101.

2. Lucanio and Colville, *American Space Science Fiction Television*, 96–101.

3. Lucanio and Coville, *American Space Science Fiction Television*, 195–99; Christopher H. Sterling and John Michael Kittross, *Stay Tuned: A History of American Broadcasting*, 3rd ed. (Mahwah, NJ: Lawrence Erlbaum Associates, 2002), 370; Morleen Getz Rouse, "A History of the F. W. Ziv Radio and Television Syndication Companies: 1930–1960" (PhD diss., University of Michigan, 1976), 113. I thank Michael Baskett for bringing Rouse's dissertation to my attention.

4. Howard E. McCurdy, *Space and the American Imagination* (Washington, DC: Smithsonian Institution Press, 1997), especially chapter 2, "Making Space Flight Seem Real," 29–51.

5. Lucanio and Coville, *American Space Science Fiction Television*, 212–17, 173.

6. Lucanio and Coville, *American Space Science Fiction Television*, 81–82, 113–18.

7. Lucanio and Coville, *American Space Science Fiction Television*, 106–8.

8. Rouse, "A History of the F. W. Ziv Radio," 3, 15.

9. Rouse, "A History of the F. W. Ziv Radio," 36, 97.

10. Rouse, "A History of the F. W. Ziv Radio," 98, 113; Christopher Anderson, "Ziv Television Productions, Inc.: U.S. Production and Syndication Company," Museum of Broadcast Communication, http://www.museum.tv/eotvsection.php?entrycode=zivtelevisio (accessed January 31, 2013).

11. Rouse, "A History of the F. W. Ziv Radio," 118.

12. Ziv, as quoted in Rouse, "A History of the F. W. Ziv Radio," 120.

13. Rouse, "A History of the F. W. Ziv Radio," 197, 200; Lucanio and Coville, *American Space Science Fiction Television*, 179.

14. Sterling and Kittross, *Stay Tuned*, 374. See also James Spiller's essay in this volume.

15. Christine Becker, *It's the Pictures That Got Small: Hollywood Film Stars on 1950s Television* (Middletown, CT: Wesleyan University Press, 2008), 17. In an interview with Rouse in 1973, Ziv remembered, "Even though I'd like to downgrade the networks in anything that I say, the fact is that everyone in our organization but me, felt flattered by going network." Rouse, "A History of the F. W. Ziv Radio," 212.

16. Lucanio and Coville, *American Space Science Fiction Television*, 150; Rouse, "A History of the F. W. Ziv Radio," 221. See also John C. Fredriksen, *Men into Space* (Albany, GA: BearManor Media, 2013).

17. Lucanio and Coville, *American Space Science Fiction Television*, 147; "Moon Probe," *Men into Space*, CBS, September 30, 1959, DB 187, 16mm print, United Artists Corporation Records, Series 7.1, Ziv Television Productions, Wisconsin Center for Film and Theater Research, Wisconsin Historical Society Archives, Madison, Wisconsin. Hereafter cited as Wisconsin Historical Society.

18. Nine episodes of *West Point* were written by a young Gene Roddenberry, who got into television work as Ziv's technical advisor from the Los Angeles Police Department for *Mr. District Attorney* (1954–55). Rouse, "A History of the F. W. Ziv Radio," 162, 231, 215.

19. "Moon Probe," *Men into Space*.

20. "Moon Probe," *Men into Space*.

21. "Moon Probe," *Men into Space*.

22. "Moon Landing," *Men into Space*, CBS, October 7, 1959, DB 188, 16mm print, United Artists Corporation Records, Series 7.1, Ziv Television Productions, Wisconsin Historical Society.

23. "Moon Landing," *Men into Space*. For an examination of the distinction that pilot astronauts made between their status and that of scientist astronauts, see Matthew H. Hersch's essay in this volume.

24. Lucanio and Coville, *American Space Science Fiction Television*, 147; "Moon Probe," *Men into Space*.

25. Nicholas de Monchaux, *Spacesuit: Fashioning Apollo* (Cambridge, MA: MIT Press, 2011), 95.

26. Lucanio and Coville, *American Space Science Fiction Television*, 142. For other aerospace researchers who experimented on themselves, see Margaret Weitekamp, *Right Stuff, Wrong Sex: America's First Women in Space Program* (Baltimore: Johns Hopkins University Press, 2004), chapter 2, 27–44. Sterling and Kittross, *Stay Tuned*, 377.

27. "Sphere of No Return," *The Man and the Challenge*, NBC, September 12, 1959, 16mm film, rel. no. 1001, Ziv TV, Copyright Collection, Motion Picture and Television Reading Room, Library of Congress, Washington, DC. Hereafter cited as Library of Congress.

28. "Experiments in Terror," *The Man and the Challenge*, NBC, October 10, 1959, DB 077, 16mm print, United Artists Corporation Records, Series 7.1, Ziv Television Productions, Wisconsin Historical Society.

29. "Border to Border," *The Man and the Challenge*, NBC, October 31, 1959, DB 080, 16mm print, United Artists Corporation Records, Series 7.1, Ziv Television Productions, Wisconsin Historical Society.

30. United Artists TV Scripts, Series 7.2, box 112, *The Man and the Challenge*, Wisconsin Historical Society.

31. "The Visitors," *The Man and the Challenge*, NBC, December 26, 1959, DB 087, 16mm print, United Artists Corporation Records, Series 7.1, Ziv Television Productions, Wisconsin Historical Society.

32. "Sphere of No Return," *The Man and the Challenge*.

33. "Sphere of No Return," *The Man and the Challenge*.

34. "Sphere of No Return," *The Man and the Challenge*.

35. "Escape to Nepal," *The Man and the Challenge*, NBC, October 24, 1959, DB 079, 16mm print, United Artists Corporation Records, Series 7.1, Ziv Television Productions, Wisconsin Historical Society.

36. "Astro Female," *The Man and the Challenge*, NBC, March 26, 1960, rel. no. 1027, 16mm film, Ziv TV, Copyright Collection, Library of Congress.

37. See also Weitekamp, *Right Stuff, Wrong Sex*, 184–6.

38. "Lost Missile," *Men into Space*, CBS, November 4, 1959, DB 191, 16mm print, United Artists Corporation Records, Series 7.1, Ziv Television Productions, Wisconsin Historical Society.

39. "Moonquake," *Men into Space*, CBS, November 11, 1959, DB 192, 16mm print, United Artists Corporation Records, Series 7.1, Ziv Television Productions, Wisconsin Historical Society.

40. "First Woman on the Moon," *Men into Space*, CBS, December 16, 1959, DB 199, 16mm print, United Artists Corporation Records, Series 7.1, Ziv Television Productions, Wisconsin Historical Society.

41. "Maximum Capacity," *The Man and the Challenge*, NBC, September 19, 1959, rel. no. 1002, 16mm film, Ziv TV, Copyright Collection, Library of Congress.

42. Elaine Tyler May, *Homeward Bound: American Families in the Cold War Era* (New York: Basic Books, 1988); Ruth Rosen, *The World Split Open: How the Women's Movement Changed America* (New York: Penguin Books, 2000); Robert D. Dean, "Masculinity as Ideology: John F. Kennedy and the Domestic Politics of Foreign Policy," *Diplomatic History* 22 (Winter 1998): 29–62; Robert D. Dean, *Imperial Brotherhood: Gender and the Making of Cold War Foreign Policy* (Amherst: University of Massachusetts Press, 2001).

43. "Odds against Survival," *The Man and the Challenge*, NBC, September 26, 1959, DB 075, 16mm print, United Artists Corporation Records, Series 7.1, Ziv Television Productions, Wisconsin Historical Society Archives.

44. "Nightmare Crossing," *The Man and the Challenge*, NBC, February 6, 1960, DB 092, 16mm print, United Artists Corporation Records, Series 7.1, Ziv Television Productions, Wisconsin Historical Society Archives.

45. "Hurricane Mesa," *The Man and the Challenge*, NBC, March 19, 1960, DB 098, 16mm print, United Artists Corporation Records, Series 7.1, Ziv Television Productions, Wisconsin Historical Society Archives.

46. "Men in a Capsule," *The Man and the Challenge*, NBC, April 9, 1960, DB 101, 16mm print, United Artists Corporation Records, Series 7.1, Ziv Television Productions, Wisconsin Historical Society Archives.

47. Lucanio and Coville, *American Space Science Fiction Television*, 150.

48. *glbtq: An Encyclopedia of Gay, Lesbian, Bisexual, Transgender, and Queer Culture*, s.v. "Nader, George" (by Linda Rapp), http://www.glbtq.com/arts/nader_g.html (accessed January 31, 2013); Army Archerd, "Nader's Death Another Sad Finale to a Glamorous H'w'd Life," *Variety*, February 4, 2002, http://www.variety.com/article/VR1117860239?refcatid=2 (accessed January 31, 2013); "George Nader, 80, Actor and Sci-Fi Writer," *New York Times*, February 12, 2002, http://www.nytimes.com/2002/02/12/arts/george-nader-80-actor-and-sci-fi-writer.html (accessed January 31, 2013).

49. Italics in the original. Dean, "Masculinity as Ideology," 30; Dean, *Imperial Brotherhood*.

50. President John F. Kennedy to Vice President Lyndon B. Johnson, memorandum, April 20, 1961, John F. Kennedy Presidential Library, Boston, Massachusetts. For more, see John Logsdon, *John F. Kennedy and the Race to the Moon*, Palgrave Studies in the History of Science and Technology (New York: Palgrave Macmillan, 2010).

51. Susan Douglas has analyzed *I Dream of Jeannie* and *Bewitched* (1964–72) as allegories about women with unnatural disruptive powers that disturb male hierarchies but who ultimately return voluntarily to their appropriate roles. Susan J. Douglas, *Where the Girls Are: Growing Up Female with the Mass Media* (New York: Times Books, 1994).

52. Anderson, "Ziv Television."

53. Rouse, "A History of the F. W. Ziv Radio," 243–45.

CHAPTER 2

> "Capsules Are Swallowed": The Mythology of the Pilot in American Spaceflight
>
> Matthew H. Hersch

Today, man, with his intelligence and reason, has suddenly come to the crossroads. Some believe that the guided missile and electronically controlled space vehicles are the ultimate answers to spaceflight. The recent orbital and suborbital achievements have been spectacular and extremely important. However, man will never be satisfied in the undignified position of sitting in a nosecone, acting as a biomedical specimen.

—Jimmy Stewart, *X-15* (1961; Santa Monica, CA: MGM Home Entertainment, 2004)

At the beginning of the 1961 feature film *X-15*, a dire-sounding Jimmy Stewart, Air Force brigadier general and movie star, insists that only one kind of man should venture into space: a pilot. The space capsules, the Soviet Vostok and the American Mercury, then being launched into space would be fine for now, Stewart declares, but men were meant to reach the stars in other ways. In the wake of the first real spaceflights, *X-15* boldly declares that America's future in space will lie in craft flown by aviators, continuing a proud tradition of piloted flight that began in 1903.

X-15 is a space movie that fails as entertainment but excels as a historical document. Produced with the support of the National Aeronautics and Space Administration (NASA), the Air Force, and North American Aviation (the contractor producing the actual experimental rocket plane of the film's title), *X-15* mixes narrative tropes with an explicit message—that the only craft with which the United States should explore space are those flown pilots. The technical advisor for *X-15* was none other than test pilot Milton "Milt" Thompson, who was slated to fly an even more radical Air Force spacecraft than the X-15, the X-20 Dyna-Soar (Figure 1).[1] Thompson (along with test pilot Neil Armstrong)

FIGURE 1. Astronaut Virgil "Gus" Grissom and test pilot Milton "Milt" Thompson stand near the Paresev 1-A at NASA's Flight Research Center in Edwards, California, in 1962. The Paresev evaluated a potential piloted return capability for NASA's Project Gemini spacecraft. Courtesy of NASA.

had earlier tried to prove that humans could control a launch vehicle manually from liftoff through orbit.[2] The X-15 and its successor, the orbital X-20, would not become America's principal vehicles for exploring space, but pilots already were, and remained for years, the principal individuals who would go there (Figure 2).

In their fifty years of professional existence, American astronauts have assumed a variety of popular roles: courageous soldier, intrepid explorer, chaste

FIGURE 2. NASA test pilot and future astronaut Neil Armstrong stands beside X-15 #1 after a research flight. Courtesy of USAF/NASA.

but alluring sex symbol.[3] Of all of the public images of the American astronaut, however, none has proven more enduring than that of the astronaut as aviator. In the decades prior to the beginning of the space age, science fiction placed a variety of characters into space, including women, elderly physicists, children, and household pets. In 1959, however, NASA's new all-male, all-military, all-pilot astronaut corps inspired a new conception of the space traveler: not mad scientist or swashbuckler, but stalwart stick-and-rudder man. Inspired by the popular image of heroic combat pilots, these taciturn, technically adept flyers were as unfazed by the specter of death as they were obsessed with accomplishing their mission.

The astronauts' real-life accomplishments beginning in 1961 not only inspired and validated this public image but pushed other kinds of potential space travelers—astronomers, engineers, physicians—out of the public imagination. Compared to the nation's pilots, America's scientists seemed particularly unfit for space exploration. Rather than inaugurating a wave of popular culture about average people living and working in space, the flights of the space shuttle beginning in 1981 inspired popular culture that reinforced the allure of the space pilot, increasingly defined not as an emotionless "systems man" but as a gruff and grizzled aviator able to steer complex machines with the nudge of a control stick. This retrograde image suited the American public, which saw in it a romanticized version of NASA's spaceflight achievements and the nation's proud legacy of individualism and heroic exploration.[4]

Spaceflight before Space Pilots

The image of the astronaut as pilot has become so entrenched in the mythology of spaceflight that it may be difficult for one to imagine an era in which some of America's brightest minds doubted whether pilots would be required at all for this endeavor. Although winged spacecraft figured prominently in 1950s concept drawings for spacecraft, in both the technical literature and the popular fiction of the era, the role of the aviator in spaceflight was often secondary to that of engineers and scientists. Although many concepts for future spacecraft created by spaceflight pioneer Wernher von Braun and other visionaries involved space planes and other piloted craft, others, including the craft depicted in the 1950 film *Destination Moon* and another designed by von Braun and Willy Ley, described mostly automated spaceships that required little hands-on piloting.[5] Although a token military aviator or navigator (or naval officer in earlier works) was commonplace aboard fictional spacecraft of the 1930s, 1940s, and 1950s, crews consisting solely of such men were not. Cinematic and television space crews required astronomers to make discoveries and engineers to keep the craft in working order, as well as colorful, clashing personalities to ensure drama.

Popular interpretations placed the fictional spaceship and its crew within the narrative conventions of the midcentury American Western, or "horse opera." In the same way that frontier towns might feature everyone from sheriffs to schoolmarms, "space operas" recreated complex extraterrestrial communities filled with grotesque characters. The eclectic space travelers of movies like *Destination Moon*, *Rocketship X-M* (1950), and *Conquest of Space* (1955) and television's *Captain Video and His Video Rangers* (1949), *Tom Corbett, Space Cadet* (1950), and *Rocky Jones, Space Ranger* (1954) were distinguished less by their piloting skills than by their resemblance to stock film and television characters: heroes and heroines, villains, damsels in distress, and comedic foils of various stripes.[6] In such films, the addition of a female chemist (*Rocketship X-M*) provided much needed sexual tension, a child acted as an audience surrogate to whom plot points could be explained (*Rocky Jones*), and an elderly professor represented a know-it-all to provide exposition and technical explanations (*Rocky Jones* again).[7] Like typical war movies of the mid-twentieth century, space films like *Rocketship X-M* and *Conquest of Space* enlivened their plot (a voyage to Mars) with multiethnic crews of technicians whose personal characteristics—accents, excitability, selfishness—provide broad humor.[8]

The enormous spacecraft of these fictional works, however, bore little resemblance to the actual craft Americans were preparing to fly into space in the 1950s: vehicles so small that they would likely accommodate only a single, highly trained occupant. Who that individual should be was a question that attracted substantial Air Force interest long before NASA's creation in 1958. By that time, America's test pilots were already edging toward space above the dry lake beds of Edwards Air Force Base, California, where, for fifteen years, aviators flying under the auspices of the Army, Air Force, Navy, and the National Advisory Committee for Aeronautics had "wrung out" a series of experimental jet- and rocket-powered aircraft intended to push the known limits of speed

FIGURE 3. The Bell Aircraft and the Martin Company joined to create this proposal for the Air Force's X-20 Dyna-Soar program, which is representative of the innovative piloted spacecraft concepts developed during the late 1950s. Piloted craft like these represented the astronauts' preferred means of accessing space. Courtesy of the Glenn L. Martin Maryland Aviation Museum.

and altitude.[9] Achieving orbital flight through incremental improvement of traditional flight technologies suited the Air Force and its pilots, and it did not stretch the imagination much to suggest that in a few short years, vehicles flown by these men would conquer space (Figure 3).

That humans would explore space in a winged craft, however, was not a foregone conclusion. In 1957, a ballistic missile hurled a Soviet probe the size of beach ball into Earth orbit. The Americans moved quickly to match this feat, and by 1959, a heated competition was already underway to launch a rudimentary piloted capsule into Earth orbit using technologies already available. These craft would dispense with wings altogether, consisting only of a spherical or conical capsule launched atop a converted intercontinental ballistic missile and guided into space and back by radio and computers instead of a pilot's skilled hands. Aviators were not obvious candidates to crew such vehicles.

In late 1958, NASA's Space Task Group undertook a brief civil service recruitment effort that, if not cancelled weeks later, would have led to the hiring of technical and scientific specialists as America's first astronauts.[10] Instead, NASA, at the insistence of President Eisenhower, drew its first seven astronauts from the all-male ranks of its military test pilots. NASA did not need actual pilots to fly its Project Mercury capsules, but it wanted the reliability and discretion military test pilots could provide. America's would-be astronauts wanted the opportunity to be the first Americans in space, but not at the expense of their professional identity. Once the astronauts arrived, they made sure that NASA, the press, and the American public knew that they would always be military aviators, first and foremost.

Birth of the Space Aviator

Despite efforts by elite government image makers to define for the American public what scientific goals spaceflight would accomplish, the public ultimately proved a challenging target for manipulation. The earliest NASA promotions of its new astronauts (including NASA-sanctioned spreads in *Life* magazine) depicted them in suits and ties, but the public favored a slightly more militant image and tended to view the space race as what Lewis Mumford disparagingly called a "symbolic act of war."[11] Within NASA's astronaut corps, a military piloting culture quickly took root and bled into the public sphere through press reports about the men. Under the astronauts' influence, NASA's space "capsules" became "spacecraft," crewed only by "pilots."[12] (As Apollo 11 astronaut Michael Collins later explained, "capsules are swallowed.")[13] Members of NASA's Space Task Group considered "space pilot" among the terms for the new professional group, ultimately modifying the word "aeronaut" to create what they thought was a neologism.[14] Even after NASA introduced the word "astronaut" to popular parlance, journalists continued to use the more specific space pilot, with its connotations of a defined professional skill set.[15]

The American public, soon instructed by President Kennedy that the exploration of space implicated the survival of the free world,[16] seized upon astronauts as pilot-warriors, risking their lives in a grand international contest.[17] Depictions in print fiction, film, and television emphasized astronauts' stolid military bearing and a stoic acceptance of danger common to popular depictions of military aviators. World War II, having ended only fifteen years earlier, provided not only the pilots who made space exploration possible (including NASA astronauts John Glenn Jr. and Donald "Deke" Slayton and the first X-15 pilot awarded "astronaut wings," Air Force Major Robert White) but also a literary vocabulary with which to describe men willing to endure such risks.[18]

The emotionless, crew-cut space pilots of television's *Men into Space* (1959–60) and *The Twilight Zone* (1959–64) and earnest films like *X-15* (1961) are recognizable caricatures, not of fictional adventurers of 1950s space movies, but of the kinds of taciturn pilots with which American audiences had become familiar thanks to a steady postwar diet of combat films. Spaceflight for these unflappable aviators is less an adventure than a solemn duty and one likely to require great sacrifice. Astronauts need not fear death, however, because they, like the bomber pilots Gregory Peck's character commands in the 1949 film *Twelve O'Clock High*, are "already dead": engaged in too perilous a pursuit to even contemplate surviving.[19] Indeed, most noteworthy of the fictional, near-future Air Force astronaut Colonel Edward McCauley (played by William Lundigan) of *Men into Space* is his utter absence of emotion when he negotiates the life-threatening crises that seem to occur in each week's space voyage.[20]

Characters ignoring their own mortality were a fixture of Rod Serling's innovative television series *The Twilight Zone*. A combat veteran strongly influenced by his experiences in the Philippines during World War II, Serling wrote or produced dozens of scripts featuring military men who heroically resist an inevitable death. Not surprisingly, *Twilight Zone* featured some of the most thoughtful investigations of the astronaut psyche, finding it somewhat akin to that of a war-hardened rifleman. The lonely, mournful men packed away in Serling's spaceships are literally not long for this world. Although death is an ever-present menace, alienation is their worst enemy: spaceflight brings a separation from loved ones, human society, familiar places, and the expectations of a full, happy life. On distant planets and asteroids, astronauts can expect not warm welcomes, but privation, murder, and imprisonment at the hands of aliens.[21] In a 1959 episode entitled "And When the Sky Was Opened," the three astronauts who return to Earth after flying their X-20 feel so alienated from society that they slowly disappear, consumed by the sense that they no longer belong on Earth.[22] Like veterans grappling with the aftereffects of wartime service, Serling's spacemen have no place in the world they left behind.

Serling's astronauts—almost all of them pilots—are stubborn, practical men. Their missions require not so much a skilled hand (which they are presumed to possess) but an ability, supposedly unique to aviators, to endure stress while preserving their composure. *Twilight Zone* spacemen encounter parallel universes, nuclear wars, and worse: the astronauts in 1963's "Death Ship" soldier on in the face of incontrovertible evidence (including hallucinations of deceased loved ones and the sight of their own corpses) that they are already dead.[23] The script by Richard Matheson (based upon an earlier novelette) ended with Serling's narration distilling the psyche of the episode's astronaut hero, "a man of such indomitable will that even the two men beneath his command are not allowed, by him, to see the truth . . . that they are no longer among the living."[24] It was precisely this kind of self-control, rather than raw piloting ability, that NASA noted with favor in its real-life space crews. "Central in their personalities," NASA psychiatrists wrote of America's first astronauts, was a "striking resilience in the face of frustration."[25]

It is this resilience, an almost maniacal devotion to exploration in the face of extreme danger, that emerges most clearly from the 1961 film *X-15*. A lazy, turgid melodrama about three pilots preparing to fly North American Aviation's rocket plane into space, it lumbers along despite the direction of Richard Donner and appearances by Mary Tyler Moore and Charles Bronson, who labor in a film so padded with stock footage that it could be mistaken for a documentary.[26] Bronson plays a test pilot anxiously awaiting the first record-breaking flights of the sleek black rocket aircraft: a craft billed in the film as simultaneously revolutionary, dangerous, and necessary. Amid predictable plotlines of love, worry, and almost senseless death, Bronson and company must harness the craft's power and confront its danger, seldom asking whether the effort is worth the trouble.

Nor does the film ever ask whether the men who have been chosen for this special destiny are the right men for the job. The X-15 flies like an airplane and thus requires pilots, but the less expensive and more reliable capsules do not. The X-15 will reach space, the film promises, but to what end, other than destiny? It is a poor platform for research in any field other than flight test engineering. Visions of spaceflight popular a decade early provided a richer context for space travel, replete with exotic worlds to be charted and surveyed. By such a measure, a space program staffed only with pilots appeared a poor force to engage in such investigations. Yet efforts by NASA to add professional scientists to its astronaut ranks would prove problematic, as popular culture built around the courageous exploits of pilots found little enthusiasm for new kinds of space travelers.

Pilots or Professors?

Ungainly, with thick-framed eyeglasses and a vaguely worried expression, the spacesuit-clad character who graces the dust jacket of astronomer Brian O'Leary's 1970 memoir *The Making of an Ex-Astronaut* bears a striking resemblance to another icon of the early 1970s: film director Woody Allen. Stuffing his head into a fishbowl helmet, O'Leary appears as out of place in his own skin as Allen's later character Miles Monroe, the New York City health food store proprietor in *Sleeper* (1973) who, in one scene, dons an ill-fitting space suit after his frozen body is thawed out 200 years into the future. Although O'Leary's text describes a poised, courageous scientist and athletic family man selected in 1967 to join the elite ranks of America's astronaut corps, it is the image of the awkward, hapless "scientist-astronaut," so unsuited to spaceflight as to be an embarrassment to his nation, that flourished in both high and low culture of the period.

Although NASA tried to sell to America, during the late 1950s and early 1960s, a group of mature, clean-cut family men as pilots for its space vehicles, the public saw them as taciturn military pilots. NASA's efforts to add civilian scientists to the flight roster beginning in 1965 met with only modest success and never galvanized public enthusiasm.[27] Popular culture lampooned space scientists as oddballs, and NASA's scientist-astronauts received a cool welcome from their pilot-astronaut colleagues. This powerful group, containing some of the most elite engineers America had ever assembled, in particular reveled in its piloting pedigree, its pedestrian sensibilities, and its supposed lack of cultural sophistication.

From the earliest days of America's first human spaceflight program, Project Mercury, NASA's astronauts leveraged their public visibility to ensure their dominance over scientists who threatened their monopoly over spaceflight. Despite their success in World War II developing the atomic bomb and microwave radar (or perhaps because of it) images of the scientist in postwar America were dominated by caricatures of oddballs engaged in seemingly magical work incomprehensible to the uninitiated. In an era of astonishing scientific advance and unprecedented funding for research,[28] movies about miraculous scientific discoveries like *The Absent Minded Professor* (1961), *Son of Flubber* (1963), and *The Nutty Professor* (1963) resonated with a public that appreciated the value of science but found it, and its practitioners, as opaque as ever.[29]

During the 1960s, rocketry increasingly resembled a large research program in which teams of engineers designed craft flown by aviators who often resented the scientists' and engineers' efforts.[30] As trained test pilots, astronauts feared becoming mere passengers in vehicles designed for monkeys. (In 1957's *Curious George Gets a Medal*, a team of balding, paunchy, pipe-smoking, bow-tie-wearing, white-coated eggheads from the Museum of Science put the finishing touches on the fearless monkey's rocket right before his flight.)[31] Although not pilots themselves, space scientists convincingly claimed expertise in spaceflight and, to the astronauts, cluttered flight plans with "Larry Lightbulb experiments" that fascinated scientists at the expense of crew safety.[32]

To support the growing needs of the space program, NASA selected new groups of astronauts roughly every other year, typically choosing from a thousand or more highly qualified military pilots for a dozen or so spots. Deke Slayton and the first American in space, Alan Shepard Jr., test pilots selected as astronauts in 1959 but grounded medically in 1962 and 1964, respectively, assumed principal responsibility for assigning these astronauts to missions and enforcing the piloting ethos. By the mid-1960s, astronaut meetings were called pilot meetings, and the Astronaut Office ensured that every member of every American space crew bore the commander or pilot designation, whether they had substantial piloting responsibilities on the craft or not.[33] It was into this often unfriendly climate that NASA's scientist-astronauts were thrown.

Under pressure from the scientific community to fulfill its research objectives and broaden its astronaut ranks to civilian experts, NASA, beginning in 1965, selected the first professional scientists to join the astronaut corps. In response to its first solicitation for qualified candidates, NASA received nearly 1,500 applications, but few applicants possessed the right combination of qualifications, and only six astronauts were approved, half the number needed. Vetted by the National Academy of Sciences, most had little flying experience but would be taught to fly as part of their NASA training. A selection in 1967, intended to produce researchers to support future long-duration Earth orbital and lunar flights, relaxed the application requirements and yielded eleven new scientist-astronauts with specialties ranging from astronomy to physiology.[34] Their diverse range of scientific talents, however, did little to endear the men to the public, NASA experts, or their astronaut peers.[35]

Skepticism about the scientists extended most ardently to the pilot-astronauts already clogging the flight roster; their criticisms, sometimes half-formed and simpleminded, reeked of popular prejudices against intellectuals whose work the pilots neither respected nor understood. Until 1965, all astronauts had been active or former military personnel, and although all of the first astronauts had undergraduate training in science or engineering, many, like Slayton and Virgil "Gus" Grissom, were "small-town boys."[36] To these salt-of-the-Earth veterans, the new MDs and PhDs were soft, weak, straight out of college, and unaccustomed to making life-or-death decisions. Astronauts who regularly teased each other by mercilessly impugning each other's flying abilities singled out the new scientist-astronauts for this kind of hazing as a group. "We quickly decided that the new breed was inferior," later wrote astronaut Walter Cunningham, summarizing the opinions of his colleagues.[37]

The presence of so many scientists in the flight rotation was unnerving to the pilots precisely because it threatened both their monopoly on space travel and their public profile. NASA's mandate to expand "human knowledge of phenomena in the atmosphere and space" made scientist-astronauts the future of the human spaceflight program, and the pilots knew it.[38] Noted Cunningham, "scientist-astronauts were brought into the program as far back as 1965. It was clear even then that they would outnumber the aviators some time in the future . . . [We] worried that Congress and the public might not know the difference, or even care." Worse, the public shared the astronauts' prejudices against scientists, and so their arrival undermined the Astronaut Office's veneer of elite skill. The presence of "milquetoast academic types" in the program suggested that many more people could be astronauts than actually were.[39]

Of the scientist-astronauts NASA selected in 1965 and 1967, only Harrison "Jack" Schmitt flew on an Apollo lunar mission and then only after the cancellation of later Apollo flights led NASA's chief scientist, Homer Newell, and the National Academy of Sciences to force Slayton to fly Schmitt on Apollo 17.[40] Geologist Schmitt, a member of the 1965 Group, was an obvious choice for a seat on a lunar landing mission, but Slayton was initially unmoved, noting that a "dead geologist" would be of "no use to anyone."[41] Although respected for his hard work in supporting other Apollo missions, Schmitt fell behind in jet pilot training and became a punch line.[42] "If God had meant man to fly," other astronauts joked, "it wouldn't have been Jack Schmitt."[43] Further galling NASA's pilot-astronauts, Schmitt's seat came at the expense of Joseph "Joe" Engle, a former X-15 test pilot known to his colleagues as "a terrific stick and rudder guy."[44]

To a public inclined to lionize astronauts as flying warriors, the response to the prospect of scientists sharing space vehicles with test pilots was tepid. In American film and television of the period, astronauts came in only two forms: heroic aviators and hapless stowaways.[45] As crewmembers on any kind of vessel, scientists had long been regarded as unwelcome guests, apt to suffer from physical weakness and odd preoccupations. In maritime culture (from which many astronauts emerged), scientists, who consumed ships' stores and burdened ships' crews with experiments but could not stand watch, were an annoyance and a hazard. "In the tradition of naval service," Helen Rozwadowski writes of efforts to integrate naturalists into nineteenth century naval operations, "a philosopher afloat used to be considered as unlucky as a cat or a corpse."[46] Likewise, in popular culture of the mid-1960s, the scientist aboard a space vehicle was invariably a helpless, inscrutable figure not to be trusted with the heroic work of spaceflight.

The "reluctant" or unqualified astronaut—often a scientist—was a particularly common motif in film and television of the period; in these depictions, scientists were either literally alien (like *Star Trek*'s extraterrestrial Mr. Spock), robotic, or merely creepy.[47] The scientists in Stanley Kubrick's *2001: A Space Odyssey* (1968) are fungible, expendable nonentities who spend the film in hibernation, individually wrapped like bags of frozen peas.[48] This portrait is relatively benign; in other films, scientists are traitors. In *Fantastic Voyage* (1966), a craft is miniaturized by a shrink ray to travel not into outer space but through the human body to save a dying man from a blood clot. A member of the vessel's scientific crew, however, is a murderous spy and saboteur.[49]

As Apollo's influence waned in the mid-1970s, skewed images of villainous space scientists remained as popular as ever, surfacing in the larger stew of cynicism about government, organized science, and big business. George's Lucas's influential 1977 blockbuster *Star Wars* celebrated the exploits of an untutored farm boy who, at the urging of a robed cleric, destroys a government-owned space weapon emblematic of "big science" itself.[50] In director Ridley Scott's

Alien (1979), a corporately owned deep space vessel's lone scientist is a sinister android in disguise, secretly programmed by the home office to recover a dangerous space monster and kill any crewmember who gets in his way.[51] The qualities that made scientists who they were, arcane knowledge, a lack of transparency about their work, and loyalty to questionable goals or ideals, continued to make them suspect.

Noteworthy about public skepticism for scientists in space was the degree to which these concerns came to be shared by educated elites. Some scientists had questioned human spaceflight's aims almost immediately upon its inception,[52] but the scientist-astronaut program broadened public skepticism.[53] Critics asserted that although manned space vehicles produced a wealth of scientific data, the programs that launched them were seldom motivated by the explicit needs of the scientific community; the space program was, to its critics, a "mission looking for a science" rather than "science looking for a mission."[54] To those within the scientific community inclined to favor robotic exploration over manned spaceflight, it was no more clear that scientists should fly in space than test pilots. Even NASA's examining psychiatrists were confused by the prospect of scientist-astronauts. The psychiatrists could not figure out why a scientist would want to fly in space, and although they asked the candidates about their motivations, the psychiatrists were not sure which responses would be pathological.[55]

The Space Shuttle and the Pilot Renaissance

During America's first decade in orbit, the public face of human space exploration belonged to taciturn test pilots with little patience for Larry Lightbulb types. In 1965, however, only a few years after the first humans ventured into space, NASA grudgingly opened its astronaut ranks to academic scientists. Although many hoped that these new "science pilots" would restore NASA's reputation and return science to the forefront of human spaceflight, with each passing year they seemed a tougher sell to the American people and to fellow astronauts. Ultimately faced with declining public interest in human spaceflight during the 1970s and 1980s, NASA management increasingly embraced popular culture's celebration not of scientists but of the common man, refashioning the human spaceflight program to build support for its efforts, with mixed success. Instead, audiences gravitated to fanciful imaginings of America's first space heroes: tough, anachronistic pilots for whom spaceflight remained a daring adventure.

As the next step in fixed-wing flight, America's space shuttle was, to many of its supporters, to be the culmination of an aeronautical engineering tradition dating back to the Wright brothers.[56] Announced by President Richard Nixon and NASA Administrator James Fletcher on January 5, 1972, the shuttle orbiters would be winged space planes the size of a small jet airliner, larger

FIGURE 4. Demonstrating the importance of pilots in the Space Shuttle Program, Commander Fred Haise and Pilot C. Gordon Fullerton pose with the orbiter *Enterprise* at the Rockwell International Space Division's Orbiter Assembly Facility in Palmdale, California, in advance of the first Shuttle Approach and Landing Tests in 1977. Courtesy of NASA.

than previous space vehicles and better able to accommodate diverse activities undertaken by nontraditional crew members. Although two pilots would still be required to fly it, the shuttle could accommodate a predominantly nonpilot crew of scientists (as many as six per flight) freed from flying responsibilities and able to devote themselves fully to scientific and other pursuits (Figure 4). In fact, the relative comfort and sophistication of the Shuttle created an opportunity for relatively untrained personnel to fly in space, a fact that NASA exploited when flights of the vehicles began in 1981. NASA, struggling to galvanize enthusiasm for its new "space truck," would celebrate the shuttle as a place where everyday people, not scientists, could ride in space as "observers."[57] (One 1986 film, *SpaceCamp*, even suggested, seemingly with NASA's endorsement, that the shuttle could be flown tolerably well by children.)[58]

NASA ultimately attempted to utilize this unique capability in a formal project to recruit nonprofessional crewmembers, the Spaceflight Participant Program, hoping, as NASA Administrator James Beggs described, that "artists" and other "professional communicators" flying aboard the space shuttle, even if assigned to do no more than "tend the galley," would help the agency connect once again to a bored public.[59] NASA also raised the specter of new problems: shortly before *SpaceCamp* opened in theaters, an explosion during the launch of the *Challenger* claimed the lives of its crew, including the first private citizen chosen to fly in it, school teacher Christa McAuliffe. True space tourism, first promised in the early 1980s, would not emerge for another twenty years.

The smaller number of astronauts who would actually pilot shuttle orbiters filled with scientists and tourists, however, emerged from this period with their reputations intact, even improved. For them, the skills required to operate the new winged craft were even more directly analogous to those of traditional aviators, and although shuttle passengers were a tough sell in popular culture, shuttle pilot-astronauts appeared in film and television even before the shuttle first flew. The heroine of the 1979 James Bond film *Moonraker* is a sultry space shuttle pilot, and Bond himself takes control of the craft in the movie's thrilling finale.[60] The television iteration of the classic Buck Rogers character from American film serials of the 1930s replaced the futuristic space voyager with a space shuttle pilot-astronaut, accidentally frozen and revived 500 years in the future. Once restored to health, Gil Gerard's character in *Buck Rogers in the 25th Century* (1979–81) finds his piloting skills to be transferable, and he seamlessly transitions into Earth's future space force, even instructing its unskilled pilots in twentieth century air combat tactics.[61]

The idea of the ancient aviator thawed out and relevant once again was a timely one in 1979. The tenth anniversary the first Apollo lunar landing encouraged a period of reexamination of the early years of America's space program and burnished the reputations of NASA's legendary pilot-astronauts. Former astronauts like Walter Cunningham had made the first forays into tell-all astronaut biography, but a "civilian" writer, Tom Wolfe, produced the most potent work on America's first spacemen, *The Right Stuff* (1979). In his often searing passages, Wolfe found a voice for a traditionally taciturn community of gruff aviators who, in Wolfe's estimation, had bucked NASA's dehumanizing organizational culture to establish the role of the pilot in the American space program. The 1983 film adaptation of the book not only recast America's first astronauts as antiheroes but rewrote the early history of American spaceflight to emphasize continuities with X-plane research of the 1940s and 1950s. Gone were the studious moments where the astronaut created a managerial identity rooted in sound engineering practice: the Original Seven astronauts of Project Mercury in the film are, instead, womanizing cowboys who must battle a cabal of heartless German-born rocket scientists (obviously intended to be von Braun and his associates, but not identified in the film) for the soul of NASA. The one question that hovers over the men is not "Why am I here?" but "Who's the best pilot you ever saw?"[62]

In space films of the 1960s, the chief asset pilots bring to their space vehicles is not a set of flying skills but a calm demeanor when confronting life-threatening situations. Astronauts personified, Howard McCurdy explains, technical perfection and unflinching competence, representative of the space program as a

FIGURE 5. Commander John Young and Pilot Robert Crippen sit on the flight deck of the orbiter *Columbia* in 1980, in preparation for the spacecraft's first flight the following year. The space shuttle offered astronauts a vehicle with airplane-style controls and a design similar to that of space plane concepts of the 1950s. Courtesy of NASA.

whole.[63] This tone matched the popular understanding of NASA's real spacecraft: dangerous machines that offered relatively few opportunities for traditional piloting. Space movies of the 1980s and 1990s, however, paradoxically celebrated a more traditional kind of aviator: the grizzled flyboy whose basic flying skills save crew, craft, and mission. The space shuttle inspired and enabled such characterizations: in 1981, two flights of the orbiter *Columbia* by all-test-pilot crews (including veteran X-15 pilot Joe Engle in the left seat) demonstrated the continued importance of aviators to NASA's human spaceflight program (Figure 5).

A common trope of these later space films is the single moment, amid a confusion of often failing technologies, when, as in the fact-based *Apollo 13* (1995) and the fictional but seemingly realistic *Deep Impact* (1998) and *Space Cowboys* (2000), a pilot's brief handling of a control stick stewards a spacecraft through a moment of extreme peril. The space shuttle's airplane-like design encouraged these plots, but writers employed the same device in films about historical spacecraft and exotic future vehicles. These moments serve many functions in these films. At the most basic level, they provide audiences with an easy-to-understand conception of the astronaut's job, one that, although inaccurate, relates to the audience's personal experience.

The circumstances requiring manual piloting in these films are greatly simplified and correspond to video game challenges easily replicated on personal computers or console devices then commonplace in American life. (Indeed, anyone who has ever tapped a joystick or game pad is apt to think "I can do that.") In *Apollo 13*, an account of the real-life aborted 1970 lunar flight, astronaut James "Jim" Lovell (Tom Hanks) must keep his ungainly craft straight as an engine is fired for its perilous return to Earth.[64] In the disaster fantasy *Deep Impact*, fictional Apollo veteran Spurgeon Tanner (Robert Duvall) must set his craft down upon a comet, and in the adventure *Space Cowboys*, over-the-hill test pilot Frank Corvin (Clint Eastwood) must land his damaged orbiter—maneuvers that require these characters to handle conventional flight controls while conjuring traditional piloting skills supposedly lost among the younger generation.[65]

These exploits validate the concept of human spaceflight generally, as these moments of flying skill usually follow mechanical failures that render automated systems useless. In particular, they serve to validate the inclusion of traditionally trained pilot-astronauts in space crews, as opposed to younger scientists or other specialists lacking such skills. Purposefully (as in *Space Cowboys*) or inadvertently (as in *SpaceCamp*), these situations reinforce the concept that the traditionally recruited and trained astronauts possess better skills than newer ones. *Space Cowboys* fictionalizes John Glenn's 1998 shuttle flight, at the age of 77, by assembling a shuttle crew composed largely of aged leading men Clint Eastwood, Tommy Lee Jones, James Garner, and Donald Sutherland. Needless to say, they resolve the film's major crises after younger astronauts are all but incapacitated. In *SpaceCamp*, a graying Tom Skerritt plays fictional veteran NASA astronaut Zach Bergstrom, whose benevolent presence on the ground hovers over a group the teenage tourists accidentally shot into space after a computer failure.

Finally, in these films, an early demonstration of flying skill establishes a single character as an unambiguous hero figure: an independent-minded aviator who, through skill and sacrifice, becomes personally responsible for the safety of others. Amid the large casts of seemingly identical characters, the protagonist with a gift for flying is a breed apart and a figure worthy of the audience's continued attention. *The Right Stuff* begins by lionizing test pilot Chuck Yeager's (Sam Shepard) breaking of the sound barrier in 1947 and ends, cryptically, when Original Seven astronaut Leroy Gordon Cooper Jr. (Dennis Quaid) hints in 1962 that it was the intrepid and largely unheralded Yeager who most deserves America's accolades. (Indeed, Yeager is nearly killed testing a new jet in the film's last and most dramatic scene).

SpaceCamp's teen piloting prodigy, Kathryn Fairly (Lea Thompson), literally flies into the film (aboard a biplane), and her status as a pilot ensures a leadership role for her as she and her team of campers learn to crew an orbiter. In *Apollo 13*, NASA astronaut Thomas Mattingly's (Gary Sinise) skill at flying an Apollo ground simulator suggests that he will play a critical later role in the film despite being bumped from the mission. Fictional future astronaut Jim McConnell in the 2000 film *Mission to Mars* (Sinise again) is referred to minutes into the film as a heroic aviator who landed a crippled Block II shuttle on a previous mission. By the movie's end, McConnell will be its undisputed emotional center as he ventures out, alone, to explore a Martian civilization.⁶⁶ In *Space Cowboys*, Corvin saves his crew's shuttle orbiter with a flick of the wrist demonstrated to him earlier in the film by his more capable colleague "Hawk" Hawkins (Tommy Lee Jones). Instead of piloting the shuttle home, daring test pilot Hawk will sacrifice his life to save Corvin and the rest of humanity by single-handedly piloting a malfunctioning Russian nuclear missile platform to the Moon.⁶⁷

The brave aviators who triumph at the conclusion of each of these movies may or may not enjoy the adoration of the American people or the pleasures of a restful retirement. *Apollo 13*'s Jim Lovell and *Space Cowboys*'s Frank Corvin will return to Earth to loving wives and the knowledge that they did their jobs as well as could be expected. For Hawk Hawkins and Jim McConnell, however, space piloting will bring a more noble triumph and a more bittersweet reward: they will have transcended their own mortality to join the pantheon of legendary lost aviators.

Conclusion: Space Pilots in Aviation History

Even more than other large technological endeavors, the history of spaceflight demonstrates how significant popular opinion has been in establishing the goals of modern American technoscience. American human spaceflight, as Walter MacDougall chronicled in *The Heavens and the Earth: A Political History of the Space Age* (1985), did not emerge in a moment of calculated calm but from a "media riot" following the Soviet launch of the first artificial Earth satellite, Sputnik, in 1957.⁶⁸ Under President Dwight D. Eisenhower, NASA recruited and publicized a force of pilots to carry America's aspirations into space. Under President John F. Kennedy, NASA fixed upon a new goal: a voyage by an American man to the Moon, which would guarantee the nation's global standing and, thanks to television, make heroes of the aviators who would achieved it. With American human spaceflight acutely dependent on congressional funding, voter enthusiasm proved influential in structuring America's agenda in space.

The Air Force, Navy, and Marine Corps test pilots who had traveled to Washington, DC, in 1959 to interview for the position of astronaut were prepared to conceal much about themselves to get the job, but the fact they tried hardest to hide was their discomfort at trading their airplanes for seats in a nose cone bolted to a rocket. The space race of the 1960s was a political problem, not an aeronautical one, and the astronauts America needed were not perfect pilots but professional managers and public relations consultants. These men would later seek to establish their piloting bona fides with both NASA's engineers and more established test pilots, who looked at Project Mercury with a mixture of contempt and alarm. Meanwhile, astronauts labored to make space capsules more like the airplanes they knew best. Passengers in NASA's new space capsules would take orders; true space pilots would give them, maintaining control over their machines and their missions.

Spaceflight's role has often been one of fulfilling a prophecy: humanity's inevitable exploration of the heavens, often described in the most grandiose and messianic of terms. Airplane mythologies in turn celebrate a kind of American—the technically proficient, independent frontiersman—that strokes preconceptions about who Americans are as a people. If the pilots' hold on space movies is weakening, evidence of such a transformation is sparse. *Red Planet* (2000) struggles to enliven the exploits of a flight engineer who battles a crazed robot on the Martian surface (yet who is still mocked as a "janitor" by the ship's pilot-in-command).[69] And the 2009 British film *Moon* restored the pilot-astronaut archetype in the form of lunar explorer Sam Bell (Sam Rockwell), although it satirized the type as a kind of inhuman fabrication of the nation's space program.[70] Ultimately, neither NASA nor the popular culture it has inspired has been able to make a passenger in a spacecraft more exciting than its pilot.

Notes

1. "Space Pilot Lands a Wingless Vehicle in Test of Re-entry," *New York Times*, July 13, 1966.

2. David A. Mindell, *Digital Apollo: Human and Machine in Spaceflight* (Cambridge, MA: MIT Press, 2008), 68–73; David A. Mindell, "Human and Machine in the History of Spaceflight," in *Critical Issues in the History of Spaceflight*, ed. Steven J. Dick and Roger D. Launius (Washington, DC: NASA, 2006), 153–54.

3. See, e.g., Roger D. Launius, "Heroes in a Vacuum: The Apollo Astronaut as Cultural Icon" (paper presented at the 43rd AIAA Aerospace Sciences Meeting and Exhibit, Reno, NV, January 10–13, 2005); Vivian Sobchack, "The Virginity of Astronauts: Sex and the Science Fiction Film," in *Alien Zone: Cultural Theory and Contemporary Science Fiction Cinema*, ed. Annette Kuhn (New York: Verso, 1990), 108.

4. This work draws, in part, on my prior research and writing on this topic. See, e.g., Matthew H. Hersch, *Inventing the American Astronaut* (New York: Palgrave MacMillan, 2012); "Return of the Lost Spaceman: America's Astronauts in Popular Culture, 1959–2006," *The Journal of Popular Culture* 44 (2011): 73–92; "Spacework: Labor And Culture in America's Astronaut Corps, 1959–1979." PhD diss., University of Pennsylvania, 2010.

5. See, e.g., *Destination Moon*, DVD, directed by Irving Pichel (1950; Chatsworth, CA: Image Entertainment, 2000); Wernher von Braun, *The Mars Project* (Urbana: University of Illinois Press, 1953). But see Wernher von Braun and Cornelius Ryan, *Conquest of the Moon* (New York: Viking Press, 1953), 36–38. "The captain of the ship, on being told by his navigator that the vehicle is off course, can make the desired change by inserting a previously prepared tape into the automatic pilot" (von Braun and Ryan, *Conquest of the Moon*, 48).

6. These television programs were instantly and extremely popular with young audiences. Patrick McCray, *Keep Watching the Skies: The Story of Operation Moonwatch and the Dawn of the Space Age* (Princeton, NJ: Princeton University Press, 2008).

7. *Rocky Jones, Space Ranger*, directed by Hollingsworth Morse (Roland Reed TV Productions, 1954). See also *Frau im Mond*, DVD, directed by Fritz Lang (1929; New York, NY: Kino Video, 2004), for a similar collection of stock characters.

8. *Rocketship X-M*, DVD, directed by Kurt Neumann (1950; Chatsworth, CA: Image Entertainment, 2000); *Conquest of Space*, DVD, directed by Byron Haskin (1955; Hollywood, CA: Paramount Home Entertainment, 2004).

9. Roger E. Bilstein, *Orders of Magnitude: A History of the NACA and NASA, 1915–1990* (Washington, DC: NASA, 1989).

10. Joseph D. Atkinson and Jay M. Shafritz, *The Real Stuff: A History of NASA's Astronaut Recruitment Program* (New York: Praeger, 1985), 32–33.

11. Lewis Mumford, "No: 'A Symbolic Act of War . . . ,'" *New York Times*, July 21, 1969.

12. NASA Space Task Group, handwritten revisions to 12/30/1960 telex from Loudon Wainwright, *Life* magazine, ARC Identifier 278231, records of the National Aeronautics and Space Administration, National Archives and Records Center, College Park, MD.

13. Michael Collins, *Carrying the Fire: An Astronaut's Journeys* (New York: Farrar, 1974), 76.

14. Colin Burgess, *Selecting the Mercury Seven: The Search for America's First Astronauts*, Springer-Praxis Books in Space Exploration (New York: Springer, 2011), 30.

15. See, e.g., "Seven Pilots Picked for Satellite Trips: 7 Chosen by U. S. as Space Pilots," *New York Times*, April 7, 1959; "2 Women Seek Roles as U.S. Space Pilots," *New York Times*, July 7, 1963; "Kennedy Phones Salute to Pilot: Watches Astronaut on TV," *New York Times*, July 22, 1961; "Space Pilots Get Training on Jets: Future Astronaut Describes Simulated Re-entry," *New York Times*, November 10, 1963.

16. John F. Kennedy, "Excerpts from 'Urgent National Needs,' Speech to a Joint Session of Congress, May 25, 1961," in *Exploring the Unknown: Selected Documents in the History of the U.S. Civil Space Program*, ed. John M. Logsdon (Washington, DC: NASA, 1995), 453.

17. Howard E. McCurdy, *Space and the American Imagination* (Washington, DC: Smithsonian Institution Press, 1997), 88–91.

18. "X-15 Test Pilot Decorated as First Winged Astronaut," *New York Times*, June 4, 1963.

19. *Twelve O'Clock High*, DVD, directed by Henry King (1949; Los Angeles, CA: 20th Century Fox Home Entertainment, 2002). Although Peck's General Frank Savage eventually crumbles under the pressure, U.S. Air Force consultants insisted that his breakdown manifest in fatigue rather than awkward bursts of irrationality and emotion. Lawrence H. Suid, *Guts and Glory: The Making of the American Military Image in Film* (Lexington: University Press of Kentucky, 2002), 112.

20. See Margaret A. Weitekamp's chapter in this volume.

21. "I Shot an Arrow into the Air," *The Twilight Zone*, directed by Stuart Rosenberg, teleplay by Rod Serling, CBS, January 15, 1960; "Elegy," *The Twilight Zone*, directed by Douglas Heyes, teleplay by Charles Beaumont, CBS, February 19, 1960; "People Are Alike All Over," *The Twilight Zone*, directed by Mitchell Leisen, teleplay by Rod Serling, CBS, March 25, 1960.

22. "And When the Sky Was Opened," *The Twilight Zone*, directed by Douglas Heyes, teleplay by Rod Serling, CBS, December 11, 1959.

23. "The Parallel," *The Twilight Zone*, directed by Alan Crosland Jr., teleplay by Rod Serling, CBS, March 14, 1963; "Probe 7, Over and Out," *The Twilight Zone*, directed by Ted Post, teleplay by Rod Serling, CBS, November 29, 1963; "Death Ship," *The Twilight Zone*, directed by Don Medford, teleplay by Richard Matheson, CBS, February 7, 1963.

24. Richard Matheson, "Death Ship," in *Richard Matheson's The Twilight Zone Scripts: Volume Two*, ed. Stanley Wiater (Springfield, PA: Edge Books, 2002), 204–5.

25. Sheldon J. Korchin and George E. Ruff, "Personality Characteristics of the Mercury Astronauts," in *The Threat of Impending Disaster, Contributions to the Psychology of Stress*, ed. George H. Grosser, Henry Wechsler, and Milton Greenblatt (Cambridge, MA: MIT Press, 1964), 204–7.

26. Even devoted space enthusiasts would likely have scoffed at the film: trying to make a feature out of a mix of widescreen narrative scenes and nonwidescreen NASA stock footage, Donner merely stretched half the shots, producing a movie that is virtually unwatchable.

27. See David J. Shayler, *NASA's Scientist-Astronauts* (New York: Springer, 2006).

28. See, e.g., Daniel J. Kevles, *The Physicists: The History of a Scientific Community in Modern America* (Cambridge, MA: Harvard University Press, 1995).

29. *The Absent Minded Professor*, DVD, directed by Robert Stevenson (1961; Burbank, CA: Walt Disney Home Video, 2003); *Son of Flubber*, DVD, directed by Robert Stevenson (1963; Burbank, CA: Walt Disney Studios Home Entertainment, 2004); *The Nutty Professor*, DVD, directed by Jerry Lewis (1963; Hollywood, CA: Paramount Home Entertainment, 2004).

30. See, e.g., Loyd S. Swenson, James M. Grimwood, and Charles C. Alexander, *This New Ocean: A History of Project Mercury* (Washington, DC: NASA, 1966); Roger E. Bilstein, *Stages to Saturn: A Technological History of the Apollo/Saturn Launch Vehicles* (Washington, DC: NASA, 1980); Thomas Parke Hughes, *American Genesis: A Century of Invention and Technological Enthusiasm, 1870–1970* (New York: Penguin Books, 1990).

31. H. A. Rey, *Curious George Gets a Medal* (Boston: Houghton Mifflin, 1957).

32. Tom Wolfe, *The Right Stuff* (New York: Bantam, 2001), 300.

33. Collins, *Carrying the Fire*, 76, 80.

34. David J. Shayler, *Skylab: America's Space Station* (Chichester, UK: Praxis, 2001), 106–7, 116. The selected astronaut candidates included engineers Owen Garriott and Edward Gibson, physicians Duane Graveline and Joseph Kerwin, physicist Frank Curtis Michel, and geologist Harrison "Jack" Schmitt.

35. The 1967 Group Scientist Astronauts included physicists Joseph Allen IV and Anthony England; engineers Philip Chapman and William Lenoir; astronomers Karl Henize, Brian O'Leary, and Robert Parker; physicians Donald Holmquest, F. Story Musgrave, and William Thornton; and chemist John Llewellyn. See, generally, Shayler, *NASA's Scientist-Astronauts*.

36. Wolfe, *The Right Stuff*, 143, 150.

37. Walter Cunningham, *The All-American Boys*, rev. ed. (New York: ibooks, 2003), 285. Although a scientist as well as a test pilot, Cunningham was forced to declare his allegiance upon joining the Astronaut Office.

38. "National Aeronautics and Space Act of 1958," in *Exploring the Unknown: Selected Documents in the History of the U.S. Civil Space Program*, ed. John M. Logsdon, NASA History Series (Washington, DC: NASA, 1995), 335.

39. Cunningham, *The All-American Boys*, 284–85.

40. Donald K. Slayton and Michael Cassutt, *Deke! U.S. Manned Space: From Mercury to the Shuttle* (New York: St. Martin's Press, 1994), 271; Richard Witkin, "Scientist Expected to Be Picked for Moon Trip: Space Agency Reported Set to Name a Geologist Move Is Viewed as Attempt to Answer Criticism," *New York Times*, December 12, 1969.

41. Shayler, *Skylab: America's Space Station*, 123. Three more of the first group of scientist-astronauts would fly on the three Skylab missions to the preliminary space station, one on each mission.

42. Slayton and Cassutt, *Deke!*, 271.

43. Cunningham, *The All-American Boys*, 108.

44. Slayton and Cassutt, *Deke!*, 271.

45. Examples include the Hollywood film *The Reluctant Astronaut* (1967), as well as television's *Stowaway to the Moon* (1975) and the short-lived series *Far Out Space Nuts* (1975).

46. "Most common sailors were rather contemptuous of the scientific 'idlers,' as they called anyone who did not stand watch." Helen M. Rozwadowski, "Small World: Forging a Scientific Maritime Culture for Oceanography," *Isis* 87 (1996): 417–18.

47. See, e.g., *Stowaway to the Moon*, directed by Andrew V. McLaglen (CBS, 1975); *The Reluctant Astronaut*, DVD, directed by Edward J. Montagne Jr. (1967; Universal City, CA: Universal Studios Home

Entertainment, 2003); Lois C. Philmus, *A Funny Thing Happened on the Way to the Moon* (New York: Spartan Books, 1966).

48. *2001: A Space Odyssey*, DVD, directed by Stanley Kubrick (1968; Burbank, CA: Warner Home Video, 2011).

49. *Fantastic Voyage*, DVD, directed by Richard Fleischer (1966; Los Angeles, CA: 20th Century Fox Home Entertainment, 2007).

50. *Star Wars*, DVD, directed by George Lucas (1977; Los Angeles, CA: 20th Century Fox Home Entertainment, 2006).

51. *Alien*, DVD, directed by Ridley Scott (1979; Los Angeles, CA: 20th Century Fox Home Entertainment, 1999).

52. Alvin M. Weinberg, "Impact of Large-Scale Science on the United States," *Science* 134 (1961): 161–64.

53. Harry Schwartz, "Space Program: Behind the Triumph, Criticism of Goals," *New York Times*, August 17, 1969; Warren E. Leary, "Debate over the Shuttle Fleet's Value to Science Has Been Raging from the Beginning," *New York Times*, February 10, 2003.

54. One complaint of scientists during Apollo, for example, was that missions occurred too frequently for adequate analysis to be undertaken of the data each mission yielded. Brian O'Leary, "Topics: Science or Stunts on the Moon?" *New York Times*, April 25, 1970.

55. Patricia A. Santy, *Choosing the Right Stuff: The Psychological Selection of Astronauts and Cosmonauts* (Westport, CT: Praeger, 1994), 34–35.

56. See, e.g., Roger E. Bilstein, *Flight in America: From the Wrights to the Astronauts* (Baltimore: Johns Hopkins University Press, 1984).

57. William J. Broad, "Reusable Space 'Truck' for Orbit Experiments," *New York Times*, April 7, 1984.

58. *SpaceCamp*, DVD, directed by Harry Winer (1986; Santa Monica, CA: MGM Home Entertainment, 2004).

59. Philip M. Boffey, "NASA to Seek Observers to Fly on Shuttles," *New York Times*, December 16, 1983.

60. *Moonraker*, DVD, directed by Lewis Gilbert (1979; Santa Monica, CA: MGM Home Entertainment, 2007).

61. *Buck Rogers in the 25th Century*, directed by Glen A. Larson, NBC, 1979–1981.

62. *The Right Stuff*, DVD, directed by Philip Kaufman (1983; Burbank, CA: Warner Home Video, 2011).

63. McCurdy, *Space and the American Imagination*, 84.

64. *Apollo 13*, DVD, directed by Ron Howard (1995; Universal City, CA: Universal Studios Home Entertainment, 2006).

65. *Deep Impact*, DVD, directed by Mimi Leder (1998; Hollywood, CA: Paramount Home Entertainment, 2004); *Space Cowboys*, DVD, directed by Clint Eastwood (2000; Burbank, CA: Warner Home Video, 2010).

66. *Mission to Mars*, DVD, directed by Brian De Palma (2000; Burbank, CA: Touchstone Home Entertainment, 2002).

67. John H. Foote, *Clint Eastwood: Evolution of a Filmmaker* (Westport, CT: Praeger, 2009), 136.

68. Walter A. McDougall, *The Heavens and the Earth: A Political History of the Space Age* (New York: Basic Books, 1985), 141.

69. *Red Planet*, DVD, directed by Antony Hoffman (2000; Burbank, CA: Warner Home Video, 2010).

70. *Moon*, directed by Duncan Jones (Hollywood, CA: Sony Pictures Classics, 2009).

CHAPTER 3

Nostalgia for the Right Stuff: Astronauts and Public Anxiety about a Changing Nation

James Spiller

The United States currently suffers a scarcity of national heroes, those rare people who attract nationwide adulation for their exceptional deeds, worthy skills, and noble commitments. New York City firefighters recently enjoyed that status owing to their losses during the horrific terrorist attacks of September 11, 2001. Soldiers rose to that rank of public esteem for their service during wars the United States subsequently fought. However, as that terrible day has receded in public memory and those costly wars ground toward untidy ends, Americans have palpably lost focus on these public servants. They have also largely forgotten another cadre of special people hailed as national heroes for much of the last fifty years: America's astronauts. They currently attract little of the public acclaim they so often did, partly because their shrinking cohort has seemed dormant since the retirement of America's space shuttles. A few astronauts still ferry into space on Russian rockets and cycle through the International Space Station, but their public profile will likely remain low until the United States launches a third phase of piloted spaceflight with a new space transportation system.

Americans elevated astronauts as national heroes during their country's first two phases of human spaceflight. The initial phase came early in the space age as the National Aeronautics and Space Administration (NASA) pursued Project Mercury, a five-year effort that began in 1958 and put the first American into space in 1961. The seven Mercury astronauts became superstars overnight after their 1959 selection. Reporters dogged them, and politicians avidly glad-handed these military test pilots because they seemingly exemplified the ideal elements of the national character. These handsome individualists—all white, male, and Protestant—were quick learning, physically fit, and steely nerved. They committed themselves to the hazardous venture of blasting solo into space, and they charted a course for their lunar-landing brethren in the Apollo program. The country's second phase of human spaceflight began in 1981 when two astronauts piloted the space shuttle *Columbia* on its maiden orbital voyage. This and subsequent shuttle crews did not draw as intense public worship as the Mercury Seven. But Americans similarly ordained these diverse men and women as national heroes, models of the highly educated team players commonly valued as the fount of national competitiveness in the 1980s.

Like their predecessors, the shuttle astronauts were commonly cast in a heroic frame familiar to most Americans. Public chroniclers of the first and second phases of American spaceflight regularly anointed astronauts as pioneers of the nation's new frontier. Although these separate generations of astronauts shared that honor, they represented two different versions of the mythic American frontier hero. The Mercury astronauts embodied a frontier type popularized in the late nineteenth century by historian turned U.S. president Theodore Roosevelt. He celebrated brave, hardy, and individual trailblazers for vitalizing the nation by preparing the western frontiers for American civilization. The shuttle astronauts evinced a very different frontier type, famously celebrated by Roosevelt's academic peer Frederick Jackson Turner. Turner also hailed frontier pioneers as the source of American exceptionalism, but he did not focus on go-it-alone explorers. Turner heralded the subsequent waves of hardworking frontier settlers as the provenance of America's exceptional virtue, its liberty, democracy, and prosperity.

These noted annalists held up what they believed were critical aspects of America's frontier past as templates for a dynamic future. Roosevelt and Turner boldly looked to that future, but they worried the nation faced challenges of domestic sclerosis and international competition. They nostalgically looked to America's frontier heroes—Roosevelt to daring pathfinders and Turner to industrious homesteaders—as models to revive Americans and inspire positive national action. Such was the case during the first two phases of American spaceflight, when public observers nostalgically treated the Mercury and shuttle astronauts as frontier pioneers. They regarded the Mercury Seven as plucky frontier trailblazers who could arouse Americans to outperform the spacefaring Soviet Union. Twenty years later, they honored the supremely trained shuttle astronauts as frontier settlers who could inspire Americans to academic excellence so that the nation could shore up its global stature. These separate generations of astronauts were similarly glorified as frontier pioneers, but they exemplified different strains of the nation's frontier mythology. Those two space-age strains followed the changing character of spaceflight between 1961 and 1981, but they also reflected Americans' changing anxieties and nostalgia for national greatness.

Mercury Astronauts as Frontier Trailblazers

In the midst of its crash effort to land men on the Moon, NASA made quick sense of America's human spaceflight program in its scholastic booklet *Space:*

The New Frontier, which asserted that "the characteristic American confidence in the future," which "brought the first colonists westward across the Atlantic" and subsequent "generations westward across the continent," had prompted the nation to set off for the space frontier.[1] Young readers were probably not surprised by this assertion, for they lived in a frontier-obsessed society. Newspapers reminded readers daily of the world's dangerous frontiers dividing freedom-loving allies from hostile enemies, just as America's western lines of settlement, according to lore, had separated it from unruly Indians. Popular entertainment brimmed with those war-whooping natives and with cowboys and cavalry for decades, often fast-forwarding these stock Wild West characters into fantastic tales of celestial travel. The popular Disneyland amusement park, opened in 1955 in Anaheim, California, did so by depicting an America born and raised in the hardscrabble Frontierland and ready to take to Tomorrowland's high-tech outposts in outer space.

Disney cast its futuristic fare on fertile ground, for space travel had long been a salable theme with frontier overtones. All manner of celestial pioneers and aliens, reminiscent of frontier beasts and bloodthirsty natives, appeared in late nineteenth and twentieth century popular entertainment. Edgar Rice Burroughs's fame and fortune rested not only on his Tarzan novels, for example, but also on his bestselling stories about the space-trotting John Carter. After Carter escaped an Indian ambush by mystical transportation to Mars, his frontiersman's strength and poise carried him through epic Martian battles. The medium that brought the similarly gallant Buck Rogers and Flash Gordon to life pumped out space fantasies from the 1920s to the 1950s, when Hollywood populated the cosmos with thinly drawn monsters whose taste for killing white men and abducting fair skinned damsels smacked of boilerplate Indians.[2] A second stream of popular culture dispensed with such frontier pulp and envisioned a technically realistic process of cosmic exploration paralleling that of the New World and the American West. Jules Verne's scientifically reasoned tales of space travel fit this bill, as did many stories in popular science fiction magazines.[3] This tradition came of age in the 1950s as international rocket societies and leading missile engineers, such as German turned U.S. Army expert Wernher von Braun, depicted in technical detail an impending age of interplanetary travel. Just as caravels and covered wagons opened New World frontiers, von Braun predicted, pioneering rocketmen would prepare the way for waves of space travelers.[4] Hollywood blockbusters, best-selling magazines, and Walt Disney television introduced millions to this scenario, to the conquest of what Disney himself called America's "new frontier, the frontier of interplanetary space."[5]

By the time the United States announced plans to make history by launching the first satellite in the late 1950s, decades of popular entertainment prepared

Americans to think of space as a new frontier, even if they did not expect to pioneer it anytime soon.⁶ They certainly did not expect the Soviet Union to blaze a trail into space, so they were surprised when it did just that in October 1957 by orbiting the satellite Sputnik. Like Frederick Jackson Turner, many had believed that frontiers were the source of America's wealth and liberalism, which in turn made it uniquely equipped to pioneer new frontiers. This was especially true in an age of advanced science and technology. As presidential advisor Vannevar Bush explained in his 1945 report *Science: The Endless Frontier*, technology and societal progress more generally "depends upon a flow of new scientific knowledge."⁷ He felt that flow was strongest in the United States owing to generous federal funding for science and to the nation's frontier legacy of freedom and democracy. This liberal legacy gave the United States an unassailable lead in science and made it inconceivable that the communist Soviet Union could leap out front and pioneer that endless frontier. As Bush confidently noted, "a totalitarian state cannot compete with a free people in the advancement of science, for the dictation and dogma are contrary to the free spirit of inquiry, which is the heart's blood of scientific advance."⁸

Such confidence failed when the Soviet Union beat the United States into space. The missile that launched Sputnik undercut America's strategic advantages and put into question a key to its global leadership, that only the United States could pioneer the endless frontier of science and its technological offspring. Thus, the *New York Times* worried that people might mistakenly believe "that Moscow has taken over world leadership in science," whereas *Life* magazine warned that the "Sputniks give this old Communist swindle a new lease of plausibility."⁹ Public officials decided to put this swindle to rest with a leading U.S. space program since it would "be construed by other nations as dramatically symbolizing national capabilities and effectiveness."¹⁰ That program would generate "a public image of supremacy," a U.S. congressman declared, and impress "the peoples of the world of that reality of [American] power."¹¹

The prominent science advisor Lloyd Berkner agreed. "In our day, when few physical frontiers remain, peoples visualize space," he announced, as a challenge "that must be accepted by a great nation to demonstrate its mettle."¹² On the advice of men like Berkner, the White House and Congress created NASA in 1958 to demonstrate America's mettle with a preeminent civil space program that projected America's benevolent global leadership while providing cover for a large military and intelligence space program. The civil program did so by showcasing the nation's superior science and technology and by putting them to use, in the words of the space agency's founding legislation, for "the benefit of all mankind." NASA spokesmen and media publicists often picked up Berkner's rhetoric and popular culture's preoccupation with the frontier by depicting the U.S. space program as a pioneering endeavor. They did so because the two strands of the space frontier motif, those associated with Frederick Jackson Turner and Theodore Roosevelt, synergistically cast this costly program as true to the nation's historic character and essential to its lofty aim of engendering global freedom and prosperity.

The frontier story of liberal vitality and economic progress that Turner told captured the generally optimistic spirit of midcentury America. It also allayed Americans' bouts of anxiety after Sputnik and after the April 1961 orbit of the first man, *a Soviet man*, which the USSR called the beginning of a "new era in human progress" made possible by "the great might of socialism."[13] NASA's program was designed to answer this provocation and "demonstrate once again," in the words of the agency's first administrator, "that free men—when challenged—can rise to the heights and overcome the lead of those who build on the basis of subjugation."[14] The Turnerian vein of the space frontier motif helped make that point since Turner's legions of pioneers demonstrated liberalism's vitality, fueled national prosperity, and laid the foundations for America's scientific and technological supremacy.[15] NASA's proponents believed that the many hundreds of thousands of Americans working to put astronauts into space would do the same. They would beat back Moscow's spacefaring challenge to the United States and open a frontier so immense that it could forever more improve the human lot. Pioneering that boundless frontier demanded a supreme pitch of performance that only a free and prosperous nation could sustain, a capacity for performance exemplified best in the early space age by the seven Mercury astronauts (Figure 1).

The Mercury Seven did not resemble Turner's bands of homesteaders. They better matched the individual trappers, scouts, and explorers exalted by Theodore Roosevelt as trailblazing exemplars for his enervated peers. Worried that his wealthy countrymen had forsaken their pioneering paternity, Roosevelt urged them to assert their racial mettle and secure the nation's global leadership by shrugging off the unnerving pull of materialism and conformity. They could do so by pursuing the "strenuous life" of muscular vigor and lofty purpose exemplified by those frontier heroes. Roosevelt warned that if Americans shrank from this noble effort, "then the bolder and stronger peoples will pass us by, and will win for themselves the domination of the world."[16] A half century later many Americans worried that materialism and conformity once again threatened their global stature. When the Soviet Union boldly leaped past the United States into space with Sputnik, public leaders howled about this staggering blow to America prestige and world power. The Catholic theologian Reinhold Niebuhr echoed Roosevelt as he warned of communist "barbarians, hardy and disciplined, ready to defeat a civilization in which the very achievements of its technology have

FIGURE 1. Mercury astronauts in their spacesuits. Courtesy of NASA (NASA image L-1989-00361).

made for soft and indulgent living."[17] On this point the evangelical preacher Billy Graham concurred. Americans "are growing soft" and their "amusements and greed for money is acting as a sedative," Graham lamented, and he channeled Roosevelt by urging the nation to "toughen up" through "compulsory scientific training in all our schools, as well as compulsory physical training for all our young people."[18] Even the U.S. Chamber of Commerce president, normally atwitter about the country's wealth and business culture, admitted to a "miasma of materialism" in America. He nevertheless declared in 1960 that the United States "has by no means been drained of its virility" and was not "an old and self-satisfied society which is ripe for collapse."[19]

His faith may then have been fortified by America's surging space program, particularly by the Mercury astronauts. These virile men were anything but self-satisfied materialists. As military officers who eschewed lucrative civilian careers, they exhibited what a presidential commission then identified as vital to the nation's future, the rejection "of a purely selfish attitude—the materialistic ethic" and commitment of "time and energy directly to the solution of the nation's problems."[20] The Mercury Seven became national heroes not only because they chose spartan earnings and high national purpose but also because they modeled outstanding bravery. As novelist Tom Wolfe noted, "reporters and broadcasters dealt with [them] in tones of awe . . . the awe that one has of an impending death-defying stunt."[21] The editors of *Life* magazine used that very tone when they predicted that if the first missile-riding "pioneer" survived, "he will become the heroic symbol of a historic triumph." His possible death was yet more awe-inspiring since "one of his six remaining comrades will go next."[22] The terrifying prospect of marching one after another into the maws of death did not break these men, for they had already faced such peril for their country as test pilots and some as combat aviators. Nor did it crush their wives, exalted by *Life* as the "seven brave women behind the astronauts," graced by their faith in God, love of country, and trust in NASA's thorough preparations.[23]

Like Roosevelt's backcountry heroes, these astronauts bravely committed themselves to a dangerous calling so that they could blaze a trail into a new frontier. Their heroic stature rested as well on their rugged strength, keen intellect, and manly individualism. The Reverend Billy Graham, who worried so much about the infirm bodies and minds of American youth, had much to admire in the Mercury Seven. Already the elite of a fit and cool-headed corps of military aviators, they famously rose to the rank of Mercury astronaut after a "harrowing ordeal of physical and mental examination." *National Geographic* detailed their athletic habits that rendered them "superb physical specimens," and it documented the intense physical training under extreme gravitational forces and in uncomfortable desert and jungle environments that gave them the "aura of supermen" (Figure 2). The magazine emphasized, however, that brute strength and physical endurance were minor virtues next to "an intellect far higher than average," which allowed the Mercury astronauts to "know their project in minute technical detail." Their native intelligence, combined with cockpit know-how and formal engineering education, enabled each astronaut to comprehend the enormously complex Mercury spaceflight systems, master his assigned "area of specialization, such as the communications network or the recovery system [and] keep one another informed, debate and weigh various ideas, and make recommendations to project officials." Each astronaut was prepared for the profound trials of spaceflight, *National Geographic* explained, for "physically and mentally, this is an extraordinary man, and for years he has worn discipline like a shield."[24]

FIGURE 2. Mercury astronauts in survival training. Courtesy of NASA (NASA image S88-31375).

As a *Life* reporter who enjoyed special access to the Mercury Seven explained, this discipline entailed more than physical strength and mental agility. The astronauts had to be able "to remain cool and resourceful under pressure." Since spaceflight "would expose them to greater stresses than most pilots had ever encountered, even in combat," they needed "nerves of steel" and an "unfailing instinct for making calm, steely, split-second decisions." Their vetting indicated what six hair-raising missions into space seemed to confirm, that they were "devoid of emotional flaws which could rattle them . . . when they found themselves in a crisis." Even more than their keen minds and athletic stamina, these nerves of steel were basic to what Tom Wolfe called the "the right stuff," the essence of the Mercury astronauts' manly fiber. But the self-possession that defined manliness required something even more than imperturbable nerves. The right stuff required unwavering individualism and independent action as well. Here too, *Life* magazine explained and documented in photos, the Mercury Seven shined. Although they underwent the same grueling battery of interviews and tests, "NASA wound up with a team of seven distinctly original personalities." They cooperated as team players and publicly denied their competition with one another, but each was "an explosive bundle of personal pride [and] professional convictions" whose "driving ambition [was] to be the first among his teammates to go into space." Their manly heroism was finally established by the presumption that they "would be the masters of their own destiny."[25] This *Life* reporter refused to believe, as many elite test pilots did, that the Mercury astronauts were "spam in a can" and would simply snore their way into space atop computer-controlled missiles and in automated capsules.

The historic May 1961 suborbital flight of Alan Shepard, in which this first spacefaring American manually controlled his capsule's attitude, indicated that Mercury astronauts were far more than rocket ballast. Their manly status as astronautical pilots was proven again in February 1962 when John Glenn took control of his *Friendship 7* capsule during America's first orbital spaceflight. When the spaceship's automatic controls faltered and a faulty signal indicated a potentially deadly reentry, Glenn took over and positioned *Friendship 7* for a safe return to Earth, where he was greeted as a national hero. According to one historian, newspapers and broadcasters subsequently depicted Glenn and his Mercury kin as self-directed "helmsmen" who combined "the pioneering image of '150 years ago' with a forward-looking mastery of technological change."[26] So too did a prominent university leader, who drew this very "analogy between the problems of the pioneers of the Western Frontier and the pioneers of outer space" and who was certain that "both ventures require personal fortitude of a high order, integrity, courage and perseverance." In his mind the flight of *Friendship 7* proved that "the elaborate equipment of a Mercury Capsule is nothing without a John Glenn."[27]

Although *Time* magazine later admitted that "astronauts often seem to be interchangeable parts of a vast mechanism," it insisted that they were in fact "essentially loners" reliant on their own pluck and skills when soaring through space.[28] It was this apparent independence that invited comparisons between the Mercury astronauts and Charles Lindbergh, the premier national hero of the aeronautical age.[29] Just as Lindbergh single-handedly flew across the Atlantic in 1927, signaling that individuals could still be self-directed pioneers in an era of industrial modernity, the solo Mercury astronauts evoked a manly independence for the technocratic space age. President Richard Nixon tapped that signal in 1970, when three of the Mercury astronauts' colleagues aboard Apollo 13 turned their lunar lander into a cosmic lifeboat to survive an explosion on their mother ship. Nixon thanked these improbable survivors for "remind[ing] us in these days when we have this magnificent technocracy, that men do count, the individual does count."[30]

The heroic stature of the Mercury astronauts hinged on the gendered nostalgia that individual *men* still counted. Many Americans worried, as Roosevelt had, that young men had grown soft in body, mind, and ambition, and they saw these astronauts as paragons of what one star-struck nun called "the manly traits we wish them to portray in their adult lives." John Glenn felt a duty to inspire young men, and he met often with groups like the Boy Scouts because it "encouraged [them] to set goals and objectives."[31] NASA leaders felt the same way and scheduled Mercury astronaut "appearances at youth group meetings" even as they cut the astronauts' many public engagements.[32] The manly heroism of the Mercury Seven, which these boys were "to portray in their adult lives," was rooted in a chauvinistic culture whose proponents believed that women lacked the strength, steady nerves, and adventurous spirit necessary to pioneer new frontiers like outer space. A de facto prohibition against female astronauts reinforced this chauvinism. Since women were not allowed to be military test pilots, then a prerequisite for the astronaut corps, women had no opportunity to prove they actually had the right stuff for spaceflight. They appeared to do so in 1959–60 when the clinic that screened the Mercury astronauts determined that women could in fact handle the rigors of space travel.[33] But the national press often dismissed the idea of petticoated space travelers. Typical was the *Philadelphia Inquirer*, which admitted that a "young unmarried woman . . . may make a fine test pilot for a simulated space ship. But wait until the simulation is gone," it joked, "and with it all chance of seeing the boyfriend for an indefinite time, and watch the old pioneer spirit evaporate into space."[34]

Like ambitious women with that necessary grit and pioneer spirit, African Americans and non-Christians had little to emulate in the Mercury astronauts. The astronauts were paraded around the world as envoys of an egalitarian nation committed to delivering new frontiers of freedom and progress to all humankind. They were celebrated testaments to this civic ideal, but they were also classic Rooseveltian heroes, uniformly white and Protestant. Theodore Roosevelt had been sufficiently open-minded to welcome diverse peoples into his idealized nation, but he believed that nation had been built and sustained by the special dispensations of Teutonic pioneers. This strain of racial thinking, which privileged white Protestant men as the fount of national progress, continued into the period of Project Mercury even as racial and religious civil rights became pressing national concerns. The African American magazine *Ebony* proudly featured black women who worked for NASA as number crunchers and black men who supported the astronaut corps as medical assistants.[35] But it was not yet able to feature astronauts of color or, for that matter, of diverse faiths. The Mercury Seven were conspicuously white and, as *Life* magazine happily revealed, faithful Protestants. They may have stood for America's gathering civic idealism, but they offered a comforting reference to the many Americans who nostalgically clung to Roosevelt's Teutonic chauvinism and who hoped the United States would continue to enjoy its special dispensations even as it became a more heterodox society.

Although the Mercury Seven comfortably came off as frontier pathfinders, public nostalgia for such heroes crashed by the 1970s as that increasingly heterodox society rejected Rooseveltian chauvinism and as the cash-strapped

United States dramatically downsized its piloted spaceflight program after winning the race to the Moon. By the time that program entered its second phase in 1981 with the first flight of space shuttle *Columbia*, shuttle crews evinced little of the Mercury astronauts' manly heroism. Owing to dramatic changes in American culture, political economy, and space technology, shuttle astronauts came off instead as pioneer settlers with graduate degrees poised to stimulate national competitiveness by commercializing the space frontier.

Shuttle Astronauts as Frontier Settlers

The editors of *Time* magazine took up the familiar trope of the space frontier when they designated "America's moon pioneers" as the "indisputable Men of the Year" of 1968. The "courage, grace, and cool efficiency" of these lunar-orbiting Apollo 8 astronauts evoked the competent bravura of Theodore Roosevelt's backwoods explorers. By asserting that the "newer world opened up by the Men of the Year will surely, in time, reach far beyond the moon," the editors echoed historian Frederick Jackson Turner as well and imagined that subsequent waves of space pioneers would launch "a journey into man's future." Although they acknowledged the nation's mounting challenges made "it easy to question the wisdom of spending billions to escape the troubled planet," they clearly failed to forecast the waning fortunes of the space frontier motif, particularly the collapse of its Rooseveltian pillars.[36]

Those chauvinistic and martial pillars had been toppled by the start of the second phase of American spaceflight. For at least a decade, African Americans and women emboldened by civil rights movements and mounting egalitarianism were no longer content celebrating their supporting roles in the U.S. spaceflight program. "We have demonstrated our belief that white-skinned, crew-cut types can do just about anything to which they set their minds," an essayist accordingly wrote about the astronaut corps in 1969; "must we not demonstrate our belief that brown-skinned, black-skinned have similar capabilities?"[37] Several years later a female editorialist, who regretted "that, in space, woman is still a joke," urged her country to acknowledge that women had those capabilities as well.[38] After the press "raked NASA over the coals for its equal opportunity shortcomings" and federal officers pushed it to recruit minorities and women, the space agency did just that and announced in 1978 that six women and three black men were among the first class of astronaut trainees for the impending space shuttles.[39] Not only did these capable men and women look different from the first cohort of American astronauts, they also did not evoke their martial spirit. Public chroniclers did not ennoble this new generation of astronauts as valiant frontier conquerors, but they did honor the shuttle astronauts as national heroes, akin to Frederick Jackson Turner's railroad age homesteaders, ready to steam into space and make that boundless new hinterland revive a flagging country.

That nostalgic honor came after a decade in which the dream of pioneering spaceflight faded as Americans grew dispirited by their stricken economy, turbulent society, and endangered environment. On the day the lunar-landing Apollo 11 astronauts splashed down in July 1969, NASA Associate Administrator for Manned Space Flight George Mueller tried to resuscitate that embattled dream by pitching a space-age version of Turner's frontier storyline. Just as Turner credited vacant hinterlands for saving America from the economic torpor, social conflict, and resource exhaustion that beset geographically confined nations, Mueller extolled outer space as a safety valve from planetary confinement. He accordingly warned his fellow citizens that if they forsook "the spirit of our forefathers then will man fall back from his destiny, the mighty surge of his achievement will be lost, and the confines of this planet will destroy him."[40] Those confines were on the minds of a class of seventh graders, who begged President Nixon to reconsider budget cutbacks so that NASA could turn the lunar frontier into "a place to spread the exploding populace, to erect 'hot houses' to help feed the world."[41] Vice President Spiro Agnew struck this note in 1969 when he warned that "the nation should never turn inward, away from the opportunities and challenges of its most promising frontiers."[42] The United States would take full advantage of those opportunities, Agnew contended, if it heeded the president's Space Task Group, which proposed a post-Apollo spaceflight program entailing a fleet of space planes, an orbital station and lunar bases, and a piloted mission to Mars.[43]

NASA Administrator Thomas Paine lobbied hard for that program, and Wernher von Braun insisted it would open the economy-boosting "new frontier of space."[44] But Nixon temporarily halted this shopworn play. He presided over a hobbled economy and determined that a retrenched space program needed to turn inward and focus on "practical application" to solve the "many critical problems here on this planet."[45] Such was the reluctant counsel of the *Washington Post*, which regretted the United States would not "press on rapidly in exploring the moon, sending men to Mars, [and] building an orbiting space station."[46] Even Thomas Paine conceded in September 1970 that "with America's concerns turning increasingly inward, and with competing budgetary demands by rapidly growing social programs, the current congressional mood was for diversified and practical space goals pursued at a moderate and economical pace."[47]

Those goals determined the form and function of the space shuttles, a new space transportation system developed throughout the 1970s at a moderate and economical pace compared to the breakneck Apollo program. When the pragmatic Nixon announced his support in 1972 for a fleet of semireusable shuttles, each of which would fly a dizzying fifty missions per year, he did not summon an imminent age of frontier expansion. The president called the project "a wise national investment" that would help "reorient our national space program so that it will have even greater domestic benefits."[48] The *New York Times* acknowledged that critics would dismiss it as "another grave distortion of national priorities . . . when so much remains undone in meeting the needs of the cities, the environment and the poor." But it endorsed the shuttle program as "a major investment in the future" that will "alter the economics of space activities and provide dividends that should continue for decades to come."[49] Earthly pragmatism was clearly the temper of the times, and CBS television's "Space: A Report to the Stockholders" broadcast those mundane dividends in businesslike terms, predicting the shuttle would positively jolt the country's listless economy.[50] So did an aerospace contractor, which hawked the future shuttle as a means to "maintain a healthy economy, keep our world trade position, solve our social ills, and build a better life."[51]

The budgetary constraints that fueled this unromantic functionalism precluded more fantastic spaceflight schemes and compelled NASA's Paine, a fervent proponent of the space frontier, to resign in fall 1970. Nixon wanted someone at the helm of NASA, one historian wrote, "either in agreement with his goal of a smaller, less costly space program or a manager who would be more pliable."[52] James Fletcher appeared to be that manager, and the new NASA chief fittingly told a hometown audience in 1972 that his agency was on track with an affordable, practical program.[53] But Fletcher turned out to be a closet evangelist for the space frontier. As he grew more concerned about Americans' pinched utilitarianism and NASA's bare bones budget, he came out of that closet, declaring that "in concentrating on the 'now' problems we are forced to ask questions about the future: are we losing sight of 'the dream'?"[54] Nixon had vaguely invoked this dream when he said the space shuttle would "help transform the space frontier," after the United States regained its financial footing, "into familiar territory, easily accessible for human endeavor in the 1980's and '90's."[55] But Fletcher reversed this timeline and suggested that an immediate push into the space frontier would energize the United States and its economy. Taking a page from Frederick Jackson Turner, he dreamed that a fast-tracked shuttle program would quickly revive the country, opening a new frontier to legions of American workers and sparking a new industrial revolution in space.

This was the "high frontier" popularized by physicist Gerard O'Neill in the late 1970s. Discounting public fear of environmental decline, resource scarcity, and zero-sum economic competition, O'Neill shared Fletcher's dream that America could rescue itself from these postfrontier ailments by rapidly colonizing the space frontier. The thriving factories and solar utilities associated with his "booming frontier settlements" would relieve Earth of its most polluting industries and end a much-feared shortage of natural resources. These settlements would also wind down domestic unrest and international conflict by creating immeasurable opportunities for economic advancement. Americans would find their gilded paths on that frontier, O'Neill cheerfully explained, owing to "the availability in the space habitats of high-paying jobs, of good living conditions, and of better opportunities."[56] This space-age disciple of Frederick Jackson Turner attracted keen attention for his anti-Malthusian message. CBS television's *60 Minutes* sounded hopeful that "some serious scientists are talking about whole colonies in space" populated by "hundreds of thousands of just plain folks looking to get away from an overcrowded Earth, running short of energy, water, and clean air."[57] Hope was also the flavor of the bicentennial anniversary edition of *National Geographic*, which featured a resource-rich future on O'Neill's high frontier as a fitting corollary to the nation's pioneering past.[58] NASA's James Fletcher was so taken by the piece, titled "The Next Frontier?," that he excitedly sent a copy to the White House.[59]

The high frontier offered an alluring blueprint, in the customary mold of Turner's frontier, for Americans who longed for better times. Presidents Nixon and Gerald Ford eschewed that nostalgic blueprint for much of the 1970s, however, and allocated just enough money so that a downsized NASA could develop a downsized space shuttle as the foundation for future spaceflight. So did President Jimmy Carter, who had been warned early on that each apostle of the space frontier "offers an ersatz frontier to replace the ones we have conquered and polluted [and] reassures us in international terms. It keeps us indisputably Number 1."[60] Carter may have heeded that warning, but his 1980 Republican rival Ronald Reagan did not. Reagan became president having promised that America would be indisputably number one again, and he expressed hope that the shuttle and his proposed space station would help revive the country as it capitalized "on Americans' pioneer spirit" and rapidly developed "our next frontier: space."[61]

That space frontier motif had not yet returned as official White House and NASA vocabulary in April 1981, when space shuttle commander John Young, one of the dozen men who had walked on the Moon, "suggested that Columbia's [maiden] journey brought man a step closer to the stars." Thus, a *Newsweek* reporter speculated this was "not precisely what NASA's public-relations people wanted to hear" since they had "been promoting the shuttle in precisely the opposite way . . . as the most practical, efficient, down-to-earth space vehicle ever designed, a 'space truck' whose mission is not exploration but the exploitation of the familiar region of nearby space."[62] But Reagan's election and the arrival of America's shuttle fleet augured a new moment when the frontier motif and spacefaring optimism, safeguarded during the 1970s by the likes of

Gerard O'Neill, became familiar once again. Like the ever-optimistic Reagan, who heralded the final "experimental" shuttle mission in June 1982 as "the historical equivalent to the driving of the golden spike which completed the transcontinental railroad," many observers buoyantly expected the spaceship would help Americans, "like the pioneers of the Old West, establish the initial settlements in space that will evolve into larger, more sophisticated facilities in the next century."[63]

This space-age version of Turner's storyline became colloquial once again as NASA depicted the shuttle as a frontier workhorse and its future space station as a frontier outpost, a valuable scientific platform, gravity-free manufacturing center, and transit point for piloted missions to the moon and Mars. These depictions appeared across popular media and carried over into many new science and technology museums. The Smithsonian Institution's National Air and Space Museum, for instance, screened the 1985 IMAX film *The Dream Is Alive*, which explained that the shuttles and planned station meant that "we now know how to live and work in space." The dream of homesteading that frontier was alive, the movie implied, because "some of our children will live in space, and their children may even be born there."[64] More than half a million kids prepared for that dream and pledged "to get ready for the 21st century" by signing up for the Young Astronauts Program, a math and science education initiative endorsed by the White House in 1984. President Reagan spelled out that program's central premise when he praised its fledglings as the "generation that will move forward to harness the enormity of space."[65] This was the generation targeted by the United States Space Camp, the first of which opened in 1982 in Huntsville, Alabama, to orient kids toward the space frontier. A promotional book titled *Your Future in Space* invited them, "as part of our first real space traveling generation," to come to camp where they would "design spacecraft for interplanetary exploration, occupy space stations," and "become a vital link in the chain of people we need to take us beyond our own planet."[66]

The shuttle crews who served as role models for this spacefaring generation were not the solo daredevils of the Mercury Project, "subjected to bone-crunching lift-offs or breath taking splash-downs." They were a "new breed of astronauts," in *Time* magazine's words, teams of shuttle pilots, mission specialists, and payload specialists that initially included men and women, blacks and Jews and a person of Asian descent.[67] This novel pedigree easily applied to the brainy civilians who made up the bulk of mission and payload specialists, but even the many shuttle pilots who had been military aviators did not fit the Mercury archetype. Unlike these heroic stick-and-rudder men, shuttle pilots came of age when the high speeds that the Mercury generation introduced, Tom Wolfe explained, "had made automatic guidance systems increasingly important."[68]

The shuttle's gentler ascent and winged descent also meant they did not need to undergo the extreme gravitational conditioning of the Mercury era. Nor did they train in forbidding deserts and jungles like their Mercury and lunar landing forebears, who prepared for emergency descents in hostile environments. Since the shuttle was designed to comfortably glide to more favorable landings, astronauts could forgo these once celebrated drills as they earnestly studied their specialized tasks in NASA's veritable college campus.[69] This change was graphically illustrated in a scholastic biography of longtime astronaut John Young, which juxtaposed a picture of the athletic Apollo trainee wrapped in desert garb with a photo of the fifty-year-old shuttle commander peering through professorial reading glasses. His relatively advanced age and softened body indicated that astronauts no longer needed to be supremely fit and have nerves of steel. Nor did they require the breathtaking courage that Young exhibited in his youth, for the space shuttle came off as a routine cargo transporter that had taken the peril out of spaceflight (Figure 3).[70]

The astronauts who flew this supposedly workaday spaceship faced the challenges of a forbidding environment with more civilized amenities than their trailblazing predecessors. In this respect they were like pioneering homesteaders, who carried their household wares via wagon and rail to hardscrabble frontiers. As shuttle astronaut Joe Allen explained, the "astronauts of the inaugural era of space travel were explorers, pioneers who ventured briefly out to the fringe of a limitless frontier." Whereas they tended "to resemble cowboys, sitting

FIGURE 3. Space Shuttle *Columbia* Commander John Young, STS-1. Courtesy of NASA (NASA image S81-30424).

in ejection seats instead of saddles, wearing helmets instead of wide-brimmed hats," Allen's colleagues were "their logical successors, men and women who go to space to work." The shuttle astronauts who did so could "perform a wider variety of intricate operations" than their cowboy predecessors and "experience more directly the sublime environment of weightless space." By ending the "Spartan era of space travel," the sizable and seemingly reliable spaceship also allowed them to feel "increasingly at home in the new frontier." Gone were the tight quarters, confining spacesuits, and squeeze-tube meals of the Mercury era. "The astronauts who now enter space wear sport shirts and slacks during their days in orbit," Allen wrote, and "eat shrimp cocktail and barbecued beef and sleep in private bunks."[71] NASA officials projected confidence in the shuttle's safety and prosaic accommodations by holding a nationwide competition in 1984 for the "first private citizen passenger in the history of spaceflight." The winning candidate, New Hampshire high school teacher Christa McAuliffe, was obviously not cut from the manly cloth of the trailblazing Mercury astronauts. But she played the role of space-age homesteader and compared herself to "the pioneer travelers of the Conestoga wagon days."[72] NASA picked up this Turnerian thread in the lesson plans it prepared for classes tracking McAuliffe's flight, asking students "to compare our future space settlers and pioneers to the early settlers and pioneers of America" and consider if "migrations from Earth to Space Stations and other planets will be similar to [their] migrations."[73]

These many allusions to Turner's frontier reflected widespread optimism about the nation's spacefaring future, but they also gained force from disconcerting evidence that the glitch-prone space shuttle was not the "most reliable, flexible, and cost-effective launch system in the world."[74] While NASA struggled to make the orbiter live up to that billing, the Soviet Union busily used expendable rockets to amass the space station experience necessary "to colonize space," and the European Space Agency's more affordable launcher beat out the shuttle and became, in the words of one journalist, "the world's only successful commercial rocket system." Japan's nascent commercial rocket venture was even more dismaying, the reporter averred, for Americans had "developed one new technology after another, from video-cassette recorders to machine tools to semi-conductors, only to watch the Japanese take the market from us."[75] Now it seemed Japan might corner the high-tech aerospace market as well.

These troubling affairs added to a general alarm that America was losing its competitive edge. As the National Commission on Excellence in Education argued in its 1983 report *A Nation at Risk*, the country's "once unchallenged preeminence in commerce, industry, science, and technological innovation is being overtaken by competitors throughout the world." If the United States hoped to "compete with them for international standing and markets," it had to encourage an academically ambitious generation to become the world's most educated workforce, and it needed to dominate cutting-edge industries.[76] The U.S. spaceflight program was supposed to accomplish both goals. Shuttle astronauts would facilitate the former as role models for a college-bound generation. Thus, the Young Astronauts Program turned the astronauts' celestial labors into inspiring math and science curriculum for motivated students, whereas the U.S. Space Camp encouraged its next cohort of space pioneers to follow the lead of shuttle astronauts and pursue advanced college degrees. Mission specialist Kathryn Sullivan picked up this theme, evident throughout media profiles of the shuttle astronauts, and advised those starry-eyed campers to develop "good 'tools' in mathematics, sciences, engineering, and English" so that they too could "be productive and successful on the space frontier."[77]

Staffed by these educated astronauts, the shuttle was supposed to accomplish the latter goal by giving the United States an insurmountable lead in the high-tech aerospace sector, a highly anticipated vanguard of the twenty-first century global economy. The shuttle's unexpected failure to do so flew in the face of President Reagan's 1982 directive to "maintain United States leadership in space."[78] So he appointed a National Commission on Space "to formulate a bold agenda to carry America's civilian space enterprise into the 21st century." Counting such apostles of the space frontier as Thomas Paine and Gerard O'Neill as members, the commission floated a very bold agenda indeed. Its 1986 report *Pioneering the Space Frontier* recommended that the United States "lead the exploration and development of the space frontier, advancing science, technology, and enterprise, and building institutions and systems that make accessible vast new resources and support human settlements beyond Earth orbit, from the highlands of the Moon to the plains of Mars." The report recapitulated Frederick Jackson Turner's economy-boosting frontier and proclaimed that "America can create new wealth on the space frontier to benefit the entire human community" by using its shuttles and planned station to secure a permanent foothold in space, which would lead to orbital industries and ultimately spawn extraplanetary settlements. This fantastical report attracted serious media attention for this Turnerian feint. It amortized the sky-high costs of the shuttle and space station over many decades, actuarially redefining these exorbitant projects as affordable means to boost national competitiveness as America led a long-term charge into the space frontier.[79]

As *Pioneering the Space Frontier* was readied for publication, a disaster occurred that undermined its rosy assumptions. When the Space Shuttle *Challenger* exploded during liftoff on January 28, 1986, and took the lives of seven astronauts, including Christa McAuliffe, Americans discovered the terrifying probability that the complex orbiters would experience catastrophic failure.

FIGURE 4. Crew of Space Shuttle *Challenger*, STS-51L. Courtesy of NASA (NASA image S85-44253).

Post-*Challenger* opinion surveys nevertheless indicated "a strong shift of public sentiment in favor of the space program generally and the shuttle program in particular," a bump in the polls that may have reflected a Rooseveltian impulse to answer the blood sacrifice of fallen frontier trailblazers by carrying on their noble endeavors.[80] Declaring "the future does not belong to the fainthearted," President Reagan channeled this indomitable impulse when he assured a grieving nation that the brave "*Challenger* crew was pulling us into the future and we'll continue to follow them" (Figure 4).[81] That bump may have also stemmed from an ongoing concern, as voiced by a corporate and university trade group, that "the U.S. lead in space is being threatened as the Soviet Union continues its ambitious space program and Europe and Japan move aggressively to harvest the bounty of space."[82] That group's Turnerian call for the United States "to develop the boundless frontier of space" found its most prominent spokesman in President George H. W. Bush, who encouraged Americans to look to outer space as "our nation's frontier, our manifest destiny." When announcing his Space Exploration Initiative in July 1989, Bush cited a familiar history, from "the voyages of Columbus to the Oregon Trail to the journey to the Moon," to galvanize support for this important national initiative, a "sustained program of manned exploration of the solar system and, yes, the permanent settlement of space."[83]

President Bush summoned the public enthusiasm evident in polls, but that devotion to the space frontier faded fast, as did his Space Exploration Initiative.

The *Challenger* accident had put the shuttle program on hold for two years, after which U.S. military priority over shuttle missions lowered its public profile. That profile dropped further when the cold war dramatically ended and Japan's economy faltered, ending the national crises that had fueled a renewed space race and sustained the space frontier motif. But the shuttle never really regained its immaculate reputation after the *Challenger* accident, and that disaster spoiled NASA's mystique for technological wizardry. Even the space agency admitted it "brought to an end over three decades of success of what was perceived to be an almost superhuman ability on the part of the American production machine to set a goal, meet it and then with equal resolve exceed that goal and set another."[84] The shuttle astronauts had personified those superhuman abilities and exhibited the academic right stuff so many people pinned as the crux of the nation's competitiveness. But as they grew more confident in America's singular power at the end of the cold war, their nostalgia for these heroic frontier pioneers dwindled, and they reveled in earthly victories rather than pioneering spaceflight as the measure of national greatness.

Nostalgia for the Right Stuff Redux

At the beginning of the first two phases of U.S. piloted spaceflight, Americans nostalgically glorified astronauts as modern-day frontier heroes. The seven Mercury astronauts came off as trailblazers whose dogged individualism and stouthearted service seemed in short supply at a time when Americans needed these traits to stay ahead of the spacefaring Soviet Union. Shuttle astronauts appeared to be space-age homesteaders whose educational achievements and specialized knowledge were generational imperatives, the foundations of national competitiveness as well as the traits needed to pioneer the space frontier. The heroic luster of the first generation of astronauts faded by the time the far more diverse and specialized shuttle crews took to orbit. The heroic stature of those crews was already in decline when a nation still aglow in post–cold war triumphalism nostalgically yearned once again for the individualist pluck and can-do risk taking associated with that first generation of astronauts. That nostalgia was on display in the 1995 blockbuster film *Apollo 13*, which celebrated a preshuttle barnstorming era of spaceflight when failure was not an option for NASA and when gutsy astronauts could still turn a near disaster into a triumphant mission. Hollywood offered an even stronger dose of rosy reminiscence in *Space Cowboys* (2000), which featured a motley mix of retired test pilots, rejected from the early astronaut corps as non–team players but called on to assist a shuttle mission. Although they were instructed to strictly follow the chain of command, these daredevils saved the endangered mission by inspiring the bureaucratic shuttle crew to improvise and rely on their own judgment.

Such fickle public adulation for America's astronauts stemmed not only from the bright spacefaring future they represented but also from the endangered traits that were so vital to the nation and so evident in the astronaut corps. The heroic stature of the next generation of astronauts will therefore depend not only on visionary goals in space but also on the earthly issues Americans find most troubling when they launch a third phase of piloted spaceflight. If the trend of the first two phases holds, those astronauts will come off once again as pioneering heroes, either trailblazing pathfinders or pioneer settlers of America's final frontier, the frontier of space.

Notes

1. *Space: The New Frontier*, EP-6 (Washington, DC: NASA, 1966), 2.

2. Brian Horrigan, "Popular Culture and Visions of the Future in Space, 1901–2001," in *New Perspectives on Technology and American Culture*, ed. Bruce Sinclair (Philadelphia: American Philosophical Society, 1986), 49–67.

3. Sam Moscowitz, "The Growth of Science Fiction from 1900 to the Early 1950s," in *Blueprint for Space: Science Fiction to Science Fact*, ed. Frederick I. Ordway III and Randy Lieberman (Washington, DC: Smithsonian Institution Press, 1992), 69–82.

4. Howard E. McCurdy, *Space and the American Imagination* (Washington, DC: Smithsonian Press, 1997), 29–51.

5. "Man in Space," ABC, March 9, 1955; *Walt Disney Treasures: Tomorrowland* (Burbank, CA: Walt Disney Studios Home Entertainment, 2004).

6. Donald N. Michael, "The Beginning of the Space Age and American Public Opinion," *Public Opinion Quarterly* 24 (1960): 573–582.

7. Vannevar Bush, *Science—The Endless Frontier* (Washington, DC: U.S. Government Printing Office, 1945), foreword.

8. Vannevar Bush, remarks at convocation of MIT, "The Essence of Security," December 5, 1949, Vannevar Bush Papers, Box 132, file "Speech, MIT 12/5/49," Library of Congress Archives, Washington, DC.

9. "Soviet Claiming Lead in Science," *The New York Times*, October 5, 1957; "A Proposal for a 'Giant Leap,'" *Life*, November 16, 1957, 53.

10. Testimony of Livingston T. Merchant to House Committee on Science and Astronautics, *Review of the Space Program*, 86th Cong., 2nd sess., 1960, 3.

11. Gretchen J. Van Dyke, "Sputnik: A Political Symbol and Tool in 1960 Campaign Politics," in *Reconsidering Sputnik: Forty Years Since the Soviet Satellite*, ed. Roger Launius, John M. Logsdon, and Robert W. Smith (London: Routledge, 2000), 396.

12. Statement by Lloyd V. Berkner, House Subcommittee on Territorial and Insular Affairs, June 10, 1960, Lloyd V. Berkner Papers, Box 21, Statement on Antarctica June 10, 1960 file, Library of Congress Manuscript Division, Washington, DC.

13. *Soviet Man in Space* (Moscow: Foreign Languages Publishing House, 1961), 12.

14. T. Keith Glennan, speech to the U.S. Air Force, August 24, 1959, Record Group 359, Subject Files 1957–62, Box 65, file "Space-Nat'l Aeronautics & Space Admin.," Executive Office of the President Office of Science and Technology, National Archives, College Park, MD.

15. Frederick Jackson Turner, "The Significance of the Frontier in American History," in *Rereading Frederick Jackson Turner*, ed. John Mack Faragher (New Haven, CT: Yale University Press, 1994), 31–32.

16. Theodore Roosevelt, "The Strenuous Life" (speech before the Hamilton Club, Chicago, April 10, 1899).

17. Reinhold Niebhur, "After Sputnik and Explorer," *Christianity and Crisis* 18, no. 4 (17 March 1958): 30.

18. Billy Graham to President Dwight D. Eisenhower, December 2, 1957, file "Impact, Public Opinion: Sputnik," Record number 006737, NASA History Office, Washington, DC.

19. Erwin D. Canham, President, Chamber of Commerce of the United States, "Business Responsibility in the World Today," keynote address, 48th Annual Meeting, May 1960. Hagley Library and Museum, 1960, Series 1, Box 21.

20. *Goals for Americans: The Report of the President's Commission on National Goals* (New York: Prentice-Hall, 1960), 23.

21. Tom Wolfe, *The Right Stuff* (New York: Bantam Books, 1979), 71.

22. Editorial, "The Astronauts—Ready to Make History," *Life*, September 14, 1959, 26.

23. Anna Glenn, "Seven Brave Women behind the Astronauts," *Life*, September 21, 1959, 142–63.

24. Allan Fisher, "Exploring Tomorrow with the Space Agency," *National Geographic*, July 1960, 52, 59.

25. John Dille, "Introduction," in *We Seven*, ed. M. Scott Carpenter, L. Gordon Cooper, Jr., John H. Glenn, Jr., Virgil I. Grissom, Walter M. Schirra, Jr., Alan B. Shepard, Jr., and Donald K. Slayton (New York: Simon and Schuster, 1962), 6–7.

26. Michael Smith, "Selling the Moon: The U.S. Manned Space Program and the Triumph of Commodity Scientism," in *The Culture of Consumption: Critical Essays in American History, 1880–1980*, ed. Richard Wightman Fox and T. J. Jackson Lears (New York: Pantheon Books, 1983), 200.

27. James Earl Rudder, "Preparation of the Whole Man," speech presented at Nuclear Space Seminar, Amarillo, Texas, August 1962, Lloyd V. Berkner Papers, Box 37, file "Space Technology and You," Library of Congress Archive, Washington, DC.

28. "On Courage in the Lunar Age," *Time*, July 25, 1969, 19.

29. Loudon S. Wainwright, "The Chosen Three for First Space Ride," *Life*, March 3, 1961, 24–33.

30. "Remarks on Presenting the Presidential Medal of Freedom to Apollo 13 Mission Operations Team in Houston, April 18, 1970," in *Public Papers of the Presidents: Richard Nixon, 1970* (Washington, DC: U.S. Government Printing Office, 1971), 367.

31. John Glenn Jr., *"P.S. I listened to your heartbeat": Letters to John Glenn* (Houston: World Book Encyclopedia Science Service, 1964), 3, 92.

32. Hiden T. Cox. memorandum, "Policy on Astronaut Appearances and Showing of MA-6 Capsule," March 9, 1962, Office of Public Affairs box, file "Correspondence," NASA History Office.

33. Margaret Weitekamp, *Right Stuff, Wrong Sex: America's First Women in Space Program* (Baltimore: Johns Hopkins University Press, 2004), 91–117.

34. "Woman Space Pioneer," *Philadelphia Inquirer*, January 15, 1959.

35. "Space Doctor for the Astronauts," *Ebony*, April 1962, 35–39; "Angel of Mercy to the Astronauts," *Ebony*, June 1966, 49–52.

36. "Men of the Year," *Time*, January 3, 1969, 1–16..

37. Robert McAfee Brown, "Moon Shot and Afterthoughts," *Christianity and Crisis* 29, no. 15 (September 15, 1969): xx.

38. Joan Ena Nauton, "A Woman in Outer Space? Good Idea—But Still a Joke," *The Miami Herald*, May 21, 1972.

39. Stan Scott to Ken Clawson, memorandum, "NASA's Image," February 19, 1974, Nixon Presidential Material Project, White House Central Files, FG 164 National Aeronautics and Space Administration, Box 2, file "[EX] Fg 164 NASA 1/1/73," National Archives, Washington, DC.

40. Statement by George Mueller, July 24, 1969, H. Guyford Stever Papers, 1930–1990, Box 66, file "NASA (1967–70)," Gerald R. Ford Library, Ann Arbor, MI.

41. Letter to President Nixon signed by 27 children, May 20, 1969, Nixon Presidential Material Project, White House Central Files, FG 164 National Aeronautics and Space Administration, Box 2, file "[Gen] NASA 3/31/70," National Archives, Washington, DC.

42. Statement by Vice President Spiro Agnew, Chairman of the National Aeronautics and Space Council, May 21, 1969. NASM, record OS-170884-50.

43. President's Space Task Force, *The Post-Apollo Space Program: Directions for the Future* (Washington, DC: U.S. Government Printing Office, 1969).

44. Wernher von Braun, "Why Is Space Exploration Vital to Man's Future," *Space World*, September 1969, 31–33.

45. "Statement about the Future of the United States Space Program," March 7, 1970, in *Public Papers of the Presidents: Richard Nixon, 1970*, 250–51.

46. "The Sun and the Planets," *Washington Post*, March 9, 1970.

47. Thomas O. Paine, "What Lies Ahead in Space," talk to the Economic Club of Detroit, September 14, 1970, file OS 170889-52, National Air and Space Museum Archives Division, Washington, DC.

48. "Annual Message to the Congress on the State of the Union," January 20, 1972, in *Public Papers of the President: Richard M. Nixon, 1972* (Washington, DC: U.S. Government Printing Office, 1973), 71.

49. "Investment in the Future," *The New York Times*, January 8, 1972.

50. "Space: A Report to Stockholder," *CBS Reports*, CBS, July 22, 1974, Library of Congress Film and Television Archives, Washington, DC.

51. *. . . for the benefit of all mankind* (Philadelphia: General Electric, 1972).

52. Roger D. Launius, "A Western Mormon in Washington, D.C.: James C. Fletcher, NASA, and the Final Frontier," *Pacific Historical Review* 64 (1995): 222.

53. Remarks by James Fletcher to the Salt Lake Rotary Club, June 6, 1972, file OS-170889-56, National Air and Space Museum Archives Division, Washington, DC.

54. *NASA and the Now Syndrome* (Washington, DC: NASA, 1975), 2.

55. "Statement Announcing Decision To Proceed With Development of the Space Shuttle," in *Public Papers of the President: Richard M. Nixon, 1972* (Washington, DC: U.S. Government Printing Office, 1973), http://www.presidency.ucsb.edu/ws/index.php?pid=3574&st=&st1 (accessed February 15, 2013).

56. Gerard O'Neill, *The High Frontier: Human Colonies in Space* (New York: Morrow, 1977), 263.

57. "Space Colonization," *60 Minutes*, CBS, October 9, 1977, Library of Congress Television and Film Archives, Washington, DC.

58. Isaac Asimov, "The Next Frontier?" *National Geographic*, July 1976, 76–89.

59. James C. Fletcher to John O. Marsh Jr., August 6, 1976, John Marsh Files, Box 109, file "Fletcher, James C.," Gerald R. Ford Library, Ann Arbor, MI.

60. Nick MacNeil to Carter-Mondale Transition Planning Group, memorandum, "NASA Recommendations," January 15, 1977, Stern Files, Box 4, file "NASA," Jimmy Carter Library, Atlanta, GA.

61. Philip M. Boffey, "President Backs U.S. Space Station as Next Key Goal," *The New York Times*, January 26, 1984.

62. Jerry Adler, John Carey, Mary Hager, and Jeff Copeland, "'In Space to Stay,'" *Newsweek*, April 27, 1981, 28.

63. Jerry Hannifin and Frederic Golden, "Space: The Once and Future Shuttle," *Time*, July 19, 1982, 72–73; George Torres, *Space Shuttle: A Quantum Leap* (Novato, CA: Presidio Press, 1986), 121.

64. *The Dream Is Alive*, directed by Graeme Ferguson (New York: IMAX Corporation, 1985).

65. Young Astronauts Council, *Young Astronaut Program Fact Sheet*, 1988; Ronald Reagan to Young Astronauts on the foundation's first anniversary, October 17, 1984, Public Papers of the President Ronald W. Reagan http://www.reagan.utexas.edu/archives/speeches/publicpapers.html (accessed February 15, 2013).

66. Flip and Debra Schulke, Penelope and Raymond McPhee, *Your Future in Space: The U.S. Space Camp Training Program* (New York: Crown, 1986), 13.

67. Frederic Golden, Sam Allis, and Jerry Hannifen, "Space: Sally Joy Ride into the Sky," *Time*, June 13, 1983, 98–102.

68. Tom Wolfe, "Columbia's Landing Closes a Circle," *National Geographic*, October 1981, 477.

69. Henry S. F. Cooper, *Before Lift-Off: The Making of a Space Shuttle Crew* (Baltimore: Johns Hopkins University Press, 1987), 15.

70. Paul Westman, *John Young: Space Shuttle Commander* (Minneapolis: Dillon Press, 1981).

71. Joseph Allen, *Entering Space: An Astronaut's Odyssey* (New York: Stewart, Tabori and Chang, 1984), 21, 31.

72. "Remarks of the Vice President Announcing the Winner of the Teacher in Space Project, July 19, 1985," in *Public Papers of the Presidents: Administration of Ronald Reagan, 1985* (Washington, DC: U.S. Government Printing Office, 1986), 932.

73. *Teacher in Space* (Washington, DC: NASA, 1985), 9.

74. *We Deliver* (Washington, DC: NASA, 1983), 2.

75. Thomas Canby, "Are the Soviets Ahead in Space," *National Geographic*, October 1986, 430; David Osbourne, "Business in Space," *The Atlantic Monthly*, May 1985, 52, 57.

76. National Commission on Excellence in Education, *A Nation at Risk: The Imperative for Educational Reform* (Washington, DC: U.S. Government Printing Office, 1983).

77. Schulke et al., *Your Future in Space*, 11.

78. "National Security Decision Directive Number 42, 'National Space Policy,' 4 July 1982," in *Exploring the Unknown: Selected Documents in the History of the U.S. Civil Space Program*, ed. John M. Logsdon, Linda J. Lear, Jannelle Warren-Findley, Ray A. Williamson, and Dwayne A. Day, (Washington, DC: NASA, 1995), 591.

79. National Commission on Space, *Pioneering the Space Frontier: The Report of the National Commission on Space* (New York: Bantam Books, 1986), 1–14.

80. Jon D. Miller, *The Impact of the Challenger Accident on Public Attitudes Toward the Space Program: A Report to the National Science Foundation* (Washington, DC: National Science Foundation, 25 January 1987), i.

81. "Address to the Nation on the Explosion of the Space Shuttle Challenger," in *Public Papers of the Presidents: Administration of Ronald Reagan, 1986* (Washington, DC: U.S. Government Printing Office, 1987).

82. *Space: America's New Competitive Frontier* (Washington, DC: Business-Higher Education Forum, April 1986), ii.

83. "Remarks on the 20th Anniversary of the *Apollo 11* Moon Landing, July 20, 1989," *Public Papers of the Presidents: Administration of George Bush, 1989* (Washington, DC: U.S. Government Printing Office, 1990), 992.

84. *Regaining the Competitive Edge* (Washington, DC: NASA, 1987).

CHAPTER 4

The Fiftieth Jubilee: Yuri Gagarin in the Soviet and Post-Soviet Imagination

Andrew Jenks

Following the collapse of the Soviet Union, the former Cosmonaut Pavilion inside the once-glorious Exhibit of People's Economic Achievements in Moscow had been converted into a flea market for gardening supplies "where foreign seeds are sold," in the words of one outraged journalist (Figure 1). An image of Gagarin in 2007 looked out over the bazaar, faded and tattered, covered by a frayed tarp as if to hide the shame of the pavilion's shabby state from the cosmonaut's gaze.[1] A graffiti-covered plane—a favorite of camera-clicking tourists—was unceremoniously dismantled in 2008, causing a group of patriotic Russians to lobby on behalf of preserving the remaining rocket in front of the pavilion and officially declaring it a cultural monument. Young Muscovites interviewed in front of the former Cosmonaut Pavilion in 2007 could remember little about Gagarin's life and feat.[2] That same year, a Moscow film company produced a movie entitled *Gagarin's Grandson* with the subtitle *A Black Comedy?* It was based on the premise that Gagarin bedded a black woman during a trip to Cameroon.[3]

Yet if one shifted the gaze from the Moscow to the Russian periphery, the attitude toward Gagarin changed dramatically. "I've never heard a joke about him," said a worker at the Engels local history museum with a straight face, adding that provincials "are more honest and open."[4] Although an annual conference in Moscow devoted to Gagarin "quietly died" in 1992, the meeting devoted to Gagarin's memory in his hometown grew in the 1990s as the "Scientific-Public Gagarin Readings." Its "scientific" mission, according to organizers, was "to preserve his sacred memory."[5] When I was interviewed for Saratov regional television in 2007, I contrasted the respect for Gagarin in Saratov with the disdainful attitudes I often encountered in Moscow. Off camera, the producer nodded approvingly, as if to say, "We know how to protect our national treasures, unlike those Muscovites."[6]

FIGURE 1. A 2007 view of the former Cosmonaut Pavilion in Moscow's Exhibit of People's Economic Achievements, turned into a flea market for gardening supplies. Photograph by the author.

The complicated relationship between the periphery and center, the engine that drove the cult of Gagarin in both the Soviet and post-Soviet eras, is the subject of this chapter. Seeking patronage from Moscow, provincial Russia participated in an exchange of sacrifices and values. The path to the stars, the provincials reminded the center, ran through places just like Klushino (Gagarin's birthplace), Saratov (where he went to school and near where he landed), and Kaluga (Konstantin Tsiolkovskii's native land). Provincials fantasized about redeeming the center with their supposed guilelessness and sincerity, embodied by the humble and youthful Gagarin, who was not spoiled by power and who had "an open and honest Russian face."[7] In exchange, the center's duty was to bestow patronage and glory back on the periphery—or at the very least to treat national heroes respectfully. It was an exchange, according to one Russian journalist, that would eliminate "the difference between past and present, between the province and the capital" and revive the nation following the humiliation of the Soviet collapse.[8] Such was the baggage attached to Yuri Gagarin's flight—a postflight mission that involved nothing less than transforming Gagarin (both before and after the collapse of the Soviet Union) into an instrument of national unification. That mission, if anything, has acquired even more importance with the passing of the fiftieth anniversary of Gagarin's flight, as Russian officials attempt to integrate aspects of a Soviet past into a serviceable post-Soviet Russian patriotism.

Premonitions of Space

In the nearly four years from Sputnik in 1957 to Gagarin's flight in 1961, Soviet citizens experienced a powerful premonition of space. With an intensity bordering on religious fervor, they anticipated the day when someone would travel to the stars. "We have been waiting for this day for so long!" exclaimed one typical unpublished letter to the editor of *Komsomol'skaia pravda*. Little wonder, then, that fulfillment of the foreseen, although supposedly secret, event provided a kind of euphoric climax. To enhance the public mood, on the day of Gagarin's flight Soviet radio broadcast *The Poem of Ecstasy*, a symphonic poem by the prerevolutionary composer Alexander Scriabin whose work, in an earlier Soviet era, had been condemned as bourgeois, decadent, mystical, and escapist.[9]

The combination of Gagarin and Scriabin projected Khrushchev era fantasies of escape from an imperfect and corrupted world. Many Soviets believed that the launching of a man into space, like the coming of Christ, presaged the dawning of a new age—as if rockets could somehow liberate people from the constraints, cramped apartments, tedium, petty arguments, boring jobs, gritty poverty, and injustices of daily life. A school teacher from a village in Tatarstan, echoing the sentiment of many of her compatriots, said the flight meant a better life was on the way: "6–7 hour work days, 2 days off a week, a happy life, and a pension when you get old." Others anticipated a more abstract payoff, "something beyond the realm of the ordinary, the realization of some kind of unclear dream." The wildly popular singer and actor Vladimir Vysotskii, a tragic figure who died young just like Gagarin, remarked on a common belief that "physics was going to unveil some ultimate secrets to humanity—and then right away cosmonauts would fly to the planets and stars, and the entire universe with its treasures and other civilizations would reveal its mysterious depths."[10]

Those dreams were so intense precisely because they drew on earlier currents in Soviet culture. Technological utopianism was a central feature of the Bolshevik world view.[11] Partly inherited from Marx's own faith in technology, it also reflected the hope that technology could solve the problem that had concerned so many of Lenin's fellow socialists: having a proletarian revolution in the most backward and impoverished country in Europe. If technological fixes were a cure-all for industrial backwardness and economic poverty, they also shaped perceptions of space flight in popular culture.[12] Long before Gagarin's flight, the potent mix of revolutionary ideology and cosmism had produced a fantasy in which space exploration would transcend the seemingly intractable problems on Earth. In the words of one scholar, "There was a widespread expectation that science, art, and technology, freed from the ties of conflicting particular interests and for the first time functioning for the benefit of humanity, would take an unprecedented upswing, pave the way for a 'bright future,' and transcend the final barrier blocking the gate to the realm of freedom—human limitations in space and time."[13] In a 1924 animated film, *Interplanetary Revolution*, the Bolsheviks vanquished the bourgeoisie of Mars and brought justice and peace to oppressed humanoids in the solar system's hinterland. It was a reaction to Iakov Protazanov's 1920s film *Aelita*, based on a short story by Aleksei Tolstoy, which fantasized about a Russian engineer who attempted to build a utopia on Mars, echoing the main theme of Aleksander Bogdanov's popular story about a Martian utopia (*Red Star*).[14] Those early works laid the foundations for a vibrant culture of science fiction reading and writing—*fantastika*, in Russian.[15]

Stalin, meanwhile, also viewed technology (along with terror) as a universal remedy, declaring that large-scale technologies, such as the Moscow metro, hulking steel factories, and massive hydroelectric dams, were preludes to a promised time of communist plenty. Space conquest, however, was more than just the latest in a long list of technological fixes. Gigantic smelters and tractor factories may have marked the early path to communism, but manned space flight, along with nuclear power, were definitely somewhere near the end of a long and bloody road that started with revolution and civil war, passed through collectivization and terror, and endured a horrific trial by fire in the war against the Nazis. And that could mean only one thing for many Soviet citizens: Gagarin was a sign that communism was imminent. "I believe in our bright future which is now just around the corner," wrote one young woman in response to Gagarin's feat, echoing the thought of a village school teacher, who was confident that "the bright future of communism will arrive precisely tomorrow." Those were typical sentiments among letter writers from the peasantry and working class.[16]

Russo-centric propaganda in the late Stalinist era had also shaped responses to Gagarin's flight. A reader of history textbooks from that era would be convinced that ethnic Russians had made practically every meaningful discovery in human history. Who else but the Great Russians—whom Stalin toasted in victory rather than the Soviets—could have defeated the Nazis? Gagarin was himself raised on fantasies of space conquest as a Russian affair that would continue the nation's record of glorious firsts, as were millions of other Soviet children. Soviet Russians were therefore ecstatic when they discovered that Gagarin was not just a Soviet citizen, but an undeniably Russian fellow, a real *muzhik* who knew what it was like to have dirt under his nails. "The Russians are in front again!" exclaimed one unpublished letter to the Soviet newspaper *Uchitel'skaia gazeta*.[17]

Blurring the boundaries between fact and fiction, science and fancy, Soviets filled Gagarin's flight with terrestrial as well as cosmic meanings. Some believed the end of war for all time was at hand, that Gagarin was "our tender dove of peace."[18] Many connected Gagarin's conquest of outer space with the placement of new virgin lands under the plow in Kazakhstan. Responding to a call on April 13 from *Komsomol'skaia pravda*, thousands of Soviets headed for the Steppe—to the "east" (*vostok*) as Gagarin's craft commanded—to create collective farms named Gagarin. Others imagined Gagarin's feat as a prelude to the invasion of new virgin lands on the Moon, Mars, and Venus, where only Russian would be spoken. "The Virgin lands of Mars are ours!" proclaimed one student. "Now we have no borders!" said another.[19] Gagarin's feat was so extraordinary that mere prose could not suffice: instead of letters to the editor, more than 300 Soviets on April 12 composed poems to the editor. Said one letter writer, prefacing his poem with mundane prose, "If you did a calculation, April 12, 1961 would be a record day for the production of poetry in the history of humankind."[20] The thoughts expressed in these and many other unpublished and published writings from ordinary Soviet citizens contained the building blocks of Soviet Russian space culture: subjugation of space (on Earth as in the cosmos), dreams of imperial occupation, messianic tendencies, and longings for eternal peace. One scholar put it this way: "As one single force—a combination of technology, fantasy, and liberation—spaceflight promised what aviation could only offer in part, total liberation from the signifiers of the past: social injustice, imperfection, gravity, and ultimately, the Earth."[21]

By April 13, Gagarin's visage was everywhere, from the ubiquitous posters featuring his dreamy gaze to the thousands of portraits dropped from planes onto the streets of Moscow and other major Soviet cities. Citizens struck a reverential pose, with their heads cocked back and eyes fixed on a cosmos filled with Soviet power. "Verticality and great height have ever been the spatial expression of potentially violent power," remarked the sociologist Henri Lefebvre.[22] Indeed, Soviets drew intense satisfaction from the thought of looking down upon their enemies. One Soviet citizen wrote to the newspaper *Literaturnaia gazeta*, "Hey Uncle Sam, listen to the rocket thunder over the entire world and over the territory of California and your various Floridas." Gagarin, wrote another, was the "first spy of the cosmos," sent on a reconnaissance mission over enemy territory. For some, Gagarin replaced the banned image of Stalin; for others he filled in for the holy icons of the Russian Orthodox tradition—a Soviet miracle worker, courtesy of Soviet technological wizardry.[23] A disabled World War II veteran clipped Gagarin's photo from the front page of *Komsomol'skaia pravda*, placed it inside a gold frame, and hung it in the corner of his home—the "beautiful corner," as Russians called it, that formerly hosted icons of Christ the Savior or Mother of God. His image, said another letter writer, was a guarantee of peace and prosperity, "like a Mauser in the hand of a sailor."[24] Still others attributed mystical significance to the length of Gagarin's flight, 108 minutes. They noted that Buddhists ring a bell 108 times for New Year's Day and that temples in Nepal have 108 niches—for 108 separate gods. The base of one of the Egyptian pyramids, some discovered, was 108 m; the speed of light in a vacuum 108×10^{10} m per hour; the speed of the Earth around the sun was 108×10^3 km per hour. "And this list could go on," noted one cosmically inclined acolyte of Gagarin's feat in the city of Engels. "108 is thus a kind of constant which the ancients lent special meaning. If Iu. Gagarin 'fit' into this number, it is because his flight 'fit' into a universal mechanism, becoming not only a scientific event but a mystical one as well."[25]

Like his flight, Gagarin's biography (from birth on an impoverished collective farm to the first man in space) followed a trajectory from a common to exalted existence. Gagarin was *nash* (one of us)—not only a "simple Soviet citizen," according to letters sent to the Academy of Sciences and Soviet newspapers, but also "our Russian man," "our Soviet Russian man," who could "endure any burden" and "master any task."[26] Said a blue-collar pensioner in an unpublished letter to the newspaper *Izvestiia*, using the informal *ty* in Russian when referring to Gagarin: "I am proud of you, a simple Soviet guy." Many letter writers followed a similar pattern, addressing Gagarin directly, rather than the editors, and offering a paean to their "simple" idol, who made them believe in the wonderful possibilities lurking somewhere within their society—and perhaps even within themselves. Some apologized: they had never written a letter to the editor before and were unsure if their grammatical skills were up to the task of praising the brave lad from Gzhatsk. But they seemed sure—after all, they sent in their letters—that Yuri, a man from the people, would understand that bad grammar in this instance was good form.[27] A man from Taganrog, in an unpublished letter, said the most important thing about Gagarin was that he was "a simple Russian guy. Let all Russian Yuris, Ivans, Mikhails, Nikolais, Andreis and the rest of the good names live eternally and grow in strength, glorifying the banner of our proletarian background." Another letter writer said Gagarin's flight made him feel a sense of kinship with the Soviet state. "It has become commonplace," wrote a medical student, "to consider public life something close to oneself."[28] Gagarin's story, in this instance, created a chain of connections that led from Gagarin to the individual and then back again to the state. Those connections allowed people to acquire, in the words of one scholar, "a sense of self as a subject and responsible agent in the world, and a recognition of the interconnection of one's own narrative with that of others."[29]

A Provincial Russian Cosmos

Like a person greeting relatives at the door, Gagarin drew Soviets, especially the Russians among them, into the Soviet state. To say that Gagarin was the first was therefore not merely to refer to his entry into space but also to highlight his status as progenitor of a new (yet paradoxically ancient) tribe/state. The heartland of this space-traveling Russian nation, as seen through the prism of the idealized Gagarin family in Soviet propaganda, was not in Moscow but in Gagarin's hometown of Gzhatsk, the provincial backwater of the Smolensk region. In the words of one ode to Gagarin entitled "The Native Side of Things," Gagarin grew up among vast "expanses of flax and thick meadows of clover," surrounded by honeybees and butterflies, yet at the center of the country. Rivers and rivulets flowed northward to Leningrad and southward "to the Mother Volga, and then meandered their way to the little father [*batiushka*] Dnieper."[30]

If people celebrated the native side of things through Gagarin, they also honored the Russian traditions that had nurtured him in that land of clover and flax (Figure 2). Beginning in the 1970s, the museum devoted to Gagarin in his hometown began researching and promoting Gagarin's childhood amusements—games such as *gorelki* (a kind of hide and seek), *lapta* (a traditional Russian

FIGURE 2. The 1973 banner of the Gagarinites in Saratov, used for parades and other official ceremonies in which the order of the Gagarinites participated. Photograph by the author.

ball and stick game), and *volchok* (like pin the tail on the donkey). By studying Gagarin's youthful diversions, children would learn a new culture of play linked to Russian folk traditions. The effort drew inspiration from a revival of Russian folk culture in the late Soviet era, when folklorists promoted traditional Russian culture as a way to solidify Russian identity and to staunch the supposedly corrosive impact of more modern and cosmopolitan cultural practices.[31] In the 1970s, children began holding "Gagarin games." The practice intensified as the Soviet Union collapsed.[32] In 1990, ethnographers, with children from Gagarin's hometown in tow, embarked on 11 expeditions to the Russian provinces to interview old timers about the games they played in their youth. The culminating point of their travels was Gagarin's birth village of Klushino, where the first cosmonaut lived as a boy until the end of World War II. Those exchanges continued in the twenty-first century, inspiring a mini-industry in ethnographic research and papers on topics such as "The Playing of Games as an Aspect of Patriotic Pedagogy in the Gagarin Family" and "Yuri Alekseevich Gagarin. The Games of a Son of the Russian Land."[33]

Those ethnographic expeditions also reenacted Gagarin's own travels after his flight. From 1961 to 1968 Gagarin crisscrossed Russia on a perpetual speaking tour, joining center to province and province to center. The rural values supposedly associated with his upbringing rooted Gagarin's travels in Russian tradition and validated it as a national path. Gagarin once remarked to a friend after he returned to Moscow from Gzhatsk that "a trip to the place where I was born . . . indeed the air itself, infused with the smells of the fields and the forests, this has filled me with new energy. Now I want to roll up my sleeves and work and study—to do what is demanded of all of us by our native land!"[34] Moscow in turn celebrated the feat of its provincial son, weighing down his chest with every conceivable medal of honor—a prelude to pilgrimage rituals that brought the center back to the supposed source of its national soul. Beginning in the 1970s, nearly 100,000 people a year boarded buses in Moscow and headed for a visit to Gagarin's ancestral homeland—a place that had, in addition, heroically endured the onslaught of both Napoleon's armies and Hitler's Wehrmacht.[35] A staple of the visit to Gagarin's birthplace of Klushino was a trip to the water well from which Gagarin drank as a boy, right next to the dugout in the ground, faithfully preserved, where he lived under Nazi occupation. Drinking deeply from that well, pilgrims connected themselves to the spirit of the family that also contained the nation's essence.

The Enduring Legacy

If provincial reverence for Gagarin drew on patriotic and spiritual motives, it also provided concrete economic benefits. World War II had essentially wiped

Gagarin's birth village of Klushino off the map. Then Gagarin's postflight glory came to the rescue. In 1961, the Soviet government created the Y. A. Gagarin Collective Farm in Klushino, employing 100 or so hardy souls. It wired the village for electricity in 1973 and in the 1970s built a museum to honor Gagarin's childhood. A few people were employed to service tourists bussed in every few days to drink from the water well from which the young Gagarin had slaked his thirst. To facilitate pilgrimages, the Soviets gave Klushino an asphalt road connection to the regional center in 1980—and even a few phone lines.[36] The village's boom years peaked in 1988, as its population nearly achieved its prerevolutionary high of about 500 souls. But then the Soviet Union collapsed, and following a pattern that had long been established, even in the Soviet years, young people left as patronage from the center dried up. In their place came migrant laborers from central Asia who did the work that young Russians refused to do in the fields. By 2004, the population had dropped by a third from 1988. Only Gagarin's memory separates Klushino, like so many other abandoned Russian rural locales, from permanent extinction or, even worse, from a fear that the native Russian stock would soon be replaced by nonnatives or natives turned into capitalist raiders and land grabbers.[37]

What was true of Klushino is equally true for the larger regional center of Gagarin, the former Gzhatsk where Gagarin moved after the war. The city reached a peak of 35,000 or so residents in the 1980s (versus 12,000 in 1940) thanks to the patronage of Gagarin and of the Soviet government. Locals played up their city's most famous son in order to attract jobs, factories, and resources for commemoration. During the 1990s, as nearly every large enterprise in the city shut down, the ambitious and able-bodied youth of Gagarin (like Gagarin himself after World War II) left to more promising lands of opportunity (Moscow was just a five-hour ride on the electric train away). One poet from the city, during a conference devoted to Gagarin's memory, wrote a poem entitled "Old Age." It recounted the story of a lonely, elderly woman from the city of Gagarin. Barely able to walk, she remarked, "All the children have left for Moscow or Smolensk. The doves have run away and soared into the sky. And what is Karmanovov or Gzhatsk to them? They only send a note on holidays." In another poem entitled "Sadness," the same author in 2007 did not mince words in describing Gagarin's dilemma. "The people are wasting away and drinking. They are unrestrained, falling down drunk. Everyone pours themselves some home-brewed hooch. We're all gulping down poison." How, he wondered, could life continue in Gagarin's homeland in the twenty-first century. "It's a tragedy . . . the whole nation lives in the dark ages. And we close our eyes. We are drowning in shit." The center had not honored its province, and so the nation seemed to be on the verge of collapse.[38] Said one local booster, "The restoration of

FIGURE 3. The main hall of the Gagarin Memorial Museum in Saratov near where Gagarin landed and, by coincidence, the place where he studied foundry work as a teenager. Photograph by the author.

historical memory—that's what we all need today. Only the preservation and restoration of places connected with our famous country-men can reestablish and strengthen the connections across time and generations."[39]

In Gagarin that warning has been heeded. Throw a stone in any direction in the city of Gagarin, and chances are it will hit a memorial space devoted to Gagarin, all under the administrative umbrella of the Y. A. Gagarin Unified Memorial Museum (Figure 3) complex.[40] Most of the museums devoted in whole or in part to Gagarin belong to the special category of "memorial museum." Honeycombing the provincial and central landscape of the Soviet Union, memorial museums were the cathedrals of a secular religion that sanctified and propagated official state values. The practice of memorializing Gagarin, however, also grew on pre-revolutionary roots, both secular and religious. "One of the main functions of the memorial museum, in contrast to other kinds of museums, is the emphasis on ritual," noted one Russian museum official at the annual Gagarin conference. The Russian Orthodox practice of worshipping saints' relics turned the environments in which they were kept into sacred space. The first secular "house" memorials were sponsored by the state and made patriotic values the object of veneration. These national ancestor cults were first associated in the eighteenth century with Peter the Great's modernization.[41] Although the Bolsheviks filled the memorial museum with their own pantheon of forebears, Lenin's tomb being the most prominent, the process of memorialization intensified after World War II. Memorial spaces devoted to war commemorated not only glorious triumphs but also, in the words of one scholar, the "patriotism of despair," a national identity based on a tale of woe and suffering, of sacrifice and devotion.[42] As Gagarin's niece put it in 2003, referring both to Russia's loss during the Great Patriotic War and her uncle's final tragic flight, "Friends recognize each other in tragedy."[43]

By the mid-1970s, almost one of every five museums in the Soviet Union received the official designation "memorial." Those museums, according to one Gagarin museum official, "marked the more important sacred moments in the history of the nation and created a material space of memory for the people. In this respect the structure of our museum network in our country differs from that in the West, where the relative percentage of memorial museums is much lower."[44] Russell Schweickart, the Apollo veteran who visited Star City in April 1983 and visited the Gagarin memorial museum inside, called the exhibit "an eye opener into Soviet sentimentality." He thought the display was "touching AND a long way from our own cultural habits."[45]

The memorial museums devoted to Gagarin in his hometown include (but are not limited to) Gagarin's childhood home (Figure 4), Gagarin's parents' house, and the hovel in the ground where Gagarin lived in the village of Klushino

FIGURE 4. A bronze statue of Gagarin's mother and her dog in yard of the Gagarin home in Gagarin. Photograph by the author.

from 1941 to 1943 under Nazi occupation (restored in 1994). The visitor encounters memorial museums within memorial museums. Local officials converted the office where Gagarin conducted correspondence when visiting his hometown after his flight into a "room museum." They encased his state-issued Volga car in glass and set it outside the Gagarin museum on Gagarin Street in Gagarin, dubbing it a "car museum." The apartment where Gagarin's mother spent her last years was converted into a memorial museum. Banners with Gagarin's helmeted head and dreamy gaze peered down on pedestrians, along with a 15 m statue of Gagarin in the city's central square. Memorial plaques graced numerous buildings and walls in Gagarin, commemorating this or that meeting Gagarin conducted as a deputy to the Supreme Soviet, all of it sacred space.[46] Even the desk where Gagarin conducted his business as a Supreme Soviet deputy when he stayed in Gzhatsk was lovingly displayed and maintained—a tribute to a time when government supposedly worked on behalf of its citizens.[47]

A Family Affair

The maintenance of sacred memory in Gagarin was both a family and national affair, which is logical since the nation was itself imagined to be Gagarin's harmonious family (Figure 5). Before she died in 1984, Gagarin's mother greeted visitors at the Gagarin museums in his home town, inviting them in for tea and chatting with them over biscuits about her son "Yura." Gagarin's sister Zoya, followed by her daughter and Gagarin's niece, later assumed that role, also plying visitors with tea and biscuits so as to communicate the "Russian spirit of hospitality" and make them "feel like family," Russian family.[48] In the 1990s, as Russia suffered the humiliating loss of great power status, they regaled visitors with tales that conveyed the spirit of the Russian peasantry, its "roots and family traditions," its "love of hard work and native Russian hospitality, [its] ability to endure stoically and bravely all of life's troubles and misfortunes," all of which allowed "a simple Russian family to produce the Columbus of the universe." In 2004 those elements were integrated into a new exhibit on Gagarin entitled "From the Family Album."[49] The memorial museum attempted to reproduce the supposed informality, intimacy, humility, and openness of Gagarin's family life and of the Russian countryside, in contrast to the "monumentality and officiousness" so often associated with urban Soviet culture. Tour guides were instructed to "note the cozy atmosphere in the house, that everything is so inviting, not only for its owners, but also for their many guests" and to emphasize Gagarin's filial piety, "his attentive relationship to his parents and his love for their home." To conjure up images of provincial suffering and sacrifice, the museum in the twenty-first century coupled Gagarin's own tragic death with the cult of Gagarin's mother, "the image of a Russian woman who had endured

FIGURE 5. The Gagarin family in Gagarin after the triumphal flight. Collection of the author.

all the difficulties of the war, who sits at home in her little peasant house, beside the oven, thinking about who did not return from the front, about all the losses." Getting Russians to contemplate loss and sacrifice—and to imagine the nation to be as hospitable and kind as the Gagarin family—such was Gagarin's task in the twenty-first century.[50]

Rarely did foreigners show up in Gagarin, and when they did, they were treated, by and large, as an oddity and even nuisance—like uninvited guests to a wake. After all, the target audience of the provincial memorial museum was Russian and not foreign. Said the director of the Unified Memorial Museum in 2003, "We have observed many times how [Russian visitors] are transformed after visiting the [Gagarin-themed] museums. They relax and become softer. Sometimes they overflow with emotions and they begin to remember their own kin. Some begin to cry. In visiting Gagarin's 'Home of Early Childhood,' visitors with great pleasure drink cold water from that same well that nurtured the Gagarins, and in doing so they connect themselves to that holy source which gives strength to a Russian person. You won't encounter that sort of thing in other exhibits."[51]

Gagarin and the Cosmist Rapture

Russianness was not the only cocktail offered at the Y. A. Gagarin Unified Memorial Museum as the Soviet Union collapsed. Within the confines of the Gagarin memorial museum the most mystical strains of cosmism had replaced communist ideology as the master narrative of Gagarin's flight.[52] Gagarin's story, according to a new scenario for the museum commissioned in 1990, provided a window into "the fate of the world," a movement of humanity toward the cosmos and into a higher state of collective consciousness.[53] The futurologist and philosopher L. V. Leskov appeared regularly at the "Gagarin" readings in Gagarin through the 1990s and early 2000s to provide more philosophical grist.[54] For Leskov, the postindustrial age began with Sputnik and Gagarin's flight. A new kind of civilization, based on a conception of world unity, peace, and the economic exploitation and colonization of space, first emerged in 1961—just as the cosmists Fedorov and Tsiokovskii had predicted. That civilization was driven forward by the restless Russian spirit, based on "the fulfillment of great ideas . . . the fate of those who have passion, of Heroic Enthusiasts." If free market ideology had led Russia into the spiritual dead end of individualism and materialism in the 1990s, Gagarin's example of passionate enthusiasm, devoid of financial interests, would return Russia to the cosmic path of glory and heroic self-sacrifice.[55]

Leskov's philosophy, which owed a debt both to the tradition of Russian cosmism and to the messianic notions of nineteenth century Slavophiles and Dostoevsky, served as the philosophical touchstone for the Unified Gagarin Memorial Museum's main post-Soviet exhibit on Gagarin entitled "The Museum of the First Flight of Man in Space."[56] The uniqueness of the exhibit, according to the director, was to devote an entire exposition to one event and then to link that event to all of world history, past, present, and future. That history, according to the exhibit's creator, was predetermined and "teleological," "an inevitable outcome placed into the evolution of the biosphere, nature, and man." By liberating himself from Earth's gravity Gagarin had also released humanity from the conventional modes of thought that made war, poverty, and suffering possible. Moreover, the universe or "biosphere" possessed a consciousness—"a mysterious intelligence"—that embodied itself in the gradual fusion of the natural and human-built world into an organic machine. And Gagarin was that being—with a beautiful soul and pitch-perfect body.[57]

No longer a socialist realist bildungsroman, Gagarin's feat had become a mystical tale of cosmic transcendence—not only in Gagarin but also in every place that celebrates his flight. "Maybe this is just fantasizing, but when you read about Gagarin and Tsiolkovskii, you can't help but fantasize," said the director of the Gagarin museum in Saratov (Figure 6). She was sure that human beings, including Gagarin, had originated from another planet. His entry into space was therefore a kind of homecoming, triggering an ancient collective memory of alien origins. "The rapture experienced when leaving this Earth, this desire to break away . . . this is the dream of all humanity," she noted. Her colleague

FIGURE 6. The monument to Gagarin at the site where he landed near the Volga city of Saratov. Photograph by the author.

in the Gagarin museum of the city of Engels across the Volga concurred. It was human nature, she said, "to want to tear away from the Earth." She suspected that Gagarin "came from there," pointing to the sky, as she explained the main theme of her 2007 Gagarin exhibit entitled "The Negation of Space." Before Gagarin, humans were "prisoners of the Earth"; afterward, they could dream of "living forever, first in orbit" and then as a kind of dispersed consciousness and intelligence in open space. Perhaps Gagarin's brain was already occupying the stellar ether, soon to be joined by future cosmists.[58]

In 2002, the Unified Gagarin Museum rearranged its relics and produced another exhibit entitled "The First Flight: World Civilization and the Russian Phenomenon," which was still on display in 2007. The exhibit attempted to answer the question "Why did the first man to enter space, preceded by all of world civilization, happen for the first time precisely in Russia?" The answer was twofold. First, it happened because of the brilliance of Russian thinkers and scientists, above all, the great Tsiolkovskii and the incomparable Korolev. But the most important factor was "the special character of the Russian people" that allowed "people from around the world to see themselves as part of a greater whole, as participants in the march of worldwide progress." The Russian spirit of self-sacrifice and martyrdom, illustrated by Russia's rescuing of the world from the evils of fascism, had propelled world history along its path of progress—just as Gagarin had.[59]

The Gagarin Art Gallery in the city of Gagarin, meanwhile, promoted its own distinctive take on the tradition of cosmism. The museum, first opened in 1984, aimed to "reveal the logic of the birth of Gagarin's heroic feat on the ancient lands of Smolensk." It quickly outgrew its original building and continued to amass Gagariniana through the post-Soviet era, commissioning regional and national artists to tap their own muses in immortalizing the first cosmonaut.[60] Gagarin's image, of course, was omnipresent—woven into fabrics, in oil, on Palekh lacquer boxes, on canvas, in bronze, in book illustrations, and in lithographs. The overarching philosophy behind the exhibit was the unity of engineering and art, science and aesthetics—just as Gagarin had embodied a fusion of man and machine. In 2007, the exhibit began with Peter the Great, who created Gagarin's hometown as a trade depot for St. Petersburg in 1719. The first hall linked images of Peter the Great and Gagarin with images of war and revolution, especially the defeat of Napoleon in 1812, when partisans from Gagarin's homeland picked off Napoleon's armies. The second hall commemorated the revolutionary movement's beginnings in the nineteenth century and the October Revolution, with images of the natural landscapes and simple people whose interests were supposedly being defended. The defeat of Nazism and postwar reconstruction "laid down a path to the first cosmonaut on Earth, Y. A. Gagarin." Along the way, much blood was spilled. "There is not a clump of earth [in Gagarin] which is not filled with people's blood," said Gagarin's mayor. "The whole of the land is built on bones and metal—that's not a political statement . . . it's the truth." The story of heroic sacrifice and martyrdom preceded the final, triumphal hall depicting Gagarin's space flight as the key historical event in the emergence of a new cosmic era. Hundreds of paintings, decorative artworks, and sculptures elucidated the Russian dream of space flight and its eventual realization, via Tsiolkovskii and Korolev, through the Russian Icarus, Gagarin.[61]

Perhaps the most striking provincial amendment to the Gagarin legend since the collapse of the Soviet Union has been his reimagination as a devout Russian Orthodox Christian. It was common to hear from Gagarin's acquaintances and Gagarin museum officials that Gagarin was secretly a believer. New legends emerged about Gagarin's efforts to save local churches from being blown up during Soviet antireligious campaigns. "He did not reject God!" proclaimed a local poet in a poem entitled "Faith." If Gagarin was a kind of Jesus, his mysterious plane crash in 1968 (at the age of 34, just like Jesus) was a test of faith, an act of sacrifice that challenged Russians to consider their sinful nature and united them in grief. Said one poet from his hometown,

> He gave his life for us
> So that people would remember and value him . . .
> Oh, how my dove of peace circles overheard!
> Over Gzhatsk he will fly for all eternity.[62]

Others pointed to something that they believed could not possibly be a coincidence: his landing on April 12 occurred at nearly the same time as Christ's Easter resurrection. It was a dramatic fulfillment of Alexander Ivanov's 1857 artistic masterpiece *Christ Appears to the People*.[63] A poem entitled "Arrival of Spring" reflected the sacred and saintly aura surrounding Gagarin in his provincial homeland:

> The star burns on our vault of heaven,
> It is there where Yura soars in the cloud dome,
> And we all await him on the horizon,
> To return to Russia, his beloved home
>
> His portrait shines like the sun,
> His sweet smile, his tender eyes they shine,
> Oh, guys, how they radiate
> Always faithful, full of honesty and virtue sublime.[64]

One religious publication in Saratov proclaimed that Gagarin had all the qualities of an Orthodox saint, "heroism, humility, greatness, charm," all of which "shone all the more after his death" and confirmed the "truth of long-forgotten Slavophile-Christian ideas." Especially saintly was his smile, which the author, a priest, said contrasted so markedly with a soulless foreigner's smile. "You'll see the distinction. Look at the young Soviet Russian officer and you'll see not only our typical broad gesture, but also a purely Christian abstract quality, an internal life, a peacefulness of the soul." The official atheism, moreover, was a trial that only made Gagarin's faith stronger. Thus, when he told Khrushchev he had not seen God in space (there is some debate as to whether he actually said this), he had not lied or violated his supposedly true faith. Instead, he was like the surgeon who once said, "I operated on the brain many times but never saw intelligence there." The intelligence came from the soul, a gift from God that had to be believed rather than seen.[65]

Booster Schemes

Soul alone, however, could not sustain the city. So Gagarin's hometown dubbed itself a "monument city," thereby hoping to attract money from rich patrons.[66] The pitch for patronage was sometimes manifested in odd and telling ways. In the mid-1990s, in the aftermath of the bloody civil war in Tadzhikistan in which ethnic Russians were targeted by nationalists and forced to flee, some contemplated turning Gagarin into a Russian refugee camp. By making Gagarin a refuge for the nation's castaways (including ethnic Russian refugees from the Chernobyl disaster) the city positioned itself as a center for gathering in the Russian refugees from the Soviet collapse. Although the plan ultimately fizzled (it envisioned factories in each of the zones of the city that would employ the forced refugees), it also reflected a harsh reality for the city and for much of provincial Russia in the 1990s: Only a refugee, with no other place to go, would stay there.[67]

But hope sprang eternal in the land of the first cosmonaut. Every year, on Gagarin's birthday, the city bigwigs staged an elaborate party, conference, and sporting competition in his hometown, sending fancy gold-embroidered invitations to VIPs from Star City, the Russian business community, and the Putin administration. According to a rite dating back to Gagarin himself, the honored guests were greeted on the border separating the Moscow and Smolensk oblasts by young maidens in folk costumes bearing the traditional bread and salt, a symbolic recreation of the unification of center and periphery.[68] During those festivities Gagarinites unveiled new memorials and monuments, as well as requests for patronage and the declamation of new odes to Gagarin. At one such ceremony in 2001 the regional administration commissioned a bigger than life bronze head of Gagarin and stuck it atop a 7 m red marble column (Figure 7). Oddly and eerily, it resembled a head on a pike, greeting passengers in the front of the train station as they exited their train cars in a semicatatonic state from the slow, rocking five-hour train ride from Moscow. City boosters then built yet another museum to honor Gagarin adjacent to the waiting hall in the train station, so that travelers "could take away something instructive and interesting" as they awaited the next leg of their journey.[69] In 2004 another project proposed new ways to market Gagarin's image—and to preserve its older identity. Although attempts to revive a moribund flax industry stalled, others proposed an annual souvenir contest to produce the best space souvenirs. After buying their fill of souvenirs, visitors would retreat to a special zone for extreme sports, inspired by Gagarin's own love of fast rockets, planes, and cars. The

FIGURE 7. A bust of Gagarin, like a head on a pike, on the train station platform of his hometown Gagarin. Photograph by the author.

theme park would be called "Star-Cruising Extreme a la Gagarin." Those plans, alas, often generated more newspaper coverage than concrete results.[70]

Booster schemes, meanwhile, continued, despite the absence of infrastructure to accommodate the anticipated tourist boom.[71] In 2006, local officials created the Gagarin Nature Park, hoping to attract ecotourists as well as hunters hoping to bag a moose, deer, or bear. Fisher people could take to the local waters for a guided fishing trip called "The Gagarin Fishing Expedition" ("guaranteed catch and tasting of the Ukho fish prepared a la Gagarin"). Tourists could also visit the Gagarin animal husbandry and fur enterprise to pet various furry creatures, local and exotic, or to buy hats, mittens, and coats made from them.[72] One local booster in 2006 proposed an alternative name for the city: "Cosmos Land." The idea was not merely to join the city of Gagarin to the space tourism and entertainment business—the most extreme of all tourist enterprises—but to make it the epicenter of a national rebirth from the ashes of empire. No other place, the boosters affirmed, could launch Russia so effectively into the cosmos, training and indoctrinating "thousands of little boys and girls from super megalopolises and villages" in the ways of Russian cosmism. Liberated from the hustle-bustle of urban capitalism, infused with a new national mission, and surrounded by forests and meadows, "the mundane affairs of everyday life will not be able to kill the young person's desire to dream, conquer, discover, and fly."[73] Visitors to Cosmos Land would be awarded a diploma and a pin testifying to his or her "candidacy as a cosmonaut."[74]

Although most of these schemes attracted more ink than money, there were signs, at least before the collapse of world financial markets in 2008 and 2009, that marketing efforts were finally beginning to pay off. "The homeland of Gagarin is gaining popularity among developers," noted a Russian newspaper joint venture of the *Wall Street Journal* and the *Financial Times*. "In the spring of 2007 in the region of the village of Klushino the company 'Gagarinland' proposed a project for the development of a land parcel of 500 hectares, on which 524 private homes are planned as well as a recreational airport with a heliport and other amenities." In 2008, a consortium of European, American, and Russian developers pledged funds of nearly $1 billion to develop tourist infrastructure in Gagarin's homeland, designed to coincide with the fiftieth anniversary of Gagarin's flight on April 12, 2011.[75]

The Memory Business

Few things in the Russian provinces have survived the Soviet era so robustly as the Gagarin cult. At least one day a year, on Cosmonautics Day, Gagarin has kept the cosmist dream of making a heroic escape from everyday life alive. Imagining themselves in space, looking down upon their problems and enemies, many Russians experience a sensation of empowerment—a typical human and not simply Russian response. "Spaciousness is closely associated with the sense of being free," noted one geographer. "Fundamental is the ability to transcend the present condition, and this transcendence is most simply manifest as the elementary power to move."[76]

Perhaps the greatest thing that Gagarin has helped provincial Russians transcend is the humiliation of the 1990s. In the places where he grew up and studied, and in many other places all around the Russian Federation, Gagarin gave Russians confidence that they could make the transition to the post-Soviet era with their national community intact, "a connecting point between the past and the future," as Gagarin's niece put it in 2007. "There was a breaking point in our lives. People were completely confused, they lost their bearings. Throwing out the old life was not so simple." Thus, even as political systems collapsed and governments changed, the annual Gagarin celebrations, wrote one journalist, "went precisely according to schedule."[77] Indeed, while people's life savings were wiped out by pyramid schemes and privatization campaigns, eager young Gagarinites received prizes for odes to their role model and for "best heroification [*geroizatsiia*]" of the first cosmonaut. Russian mothers with large families received an award in the name of Gagarin's mother for their efforts to keep the nation stocked with potential heroes.[78] Said Gagarin's friend and fellow cosmonaut Leonov in 2004 to a gathering on the central square of Gagarin for what would have been the first cosmonaut's seventieth birthday, "Everything bad in the last few years in our country is slowly disappearing, and I believe that as I look at the thousands of Gagarinites gathered on this square . . . Within our hearts there has been aroused a feeling of national pride and national self-consciousness, a realization that we are a great country." One poet at a Gagarin conference in the city of Gagarin in 2006 declaimed a composition to a packed audience entitled "The City of Gagarin: The World's Most Famous Province." Every detail of provincial life oozed "Russianness" (*Russkost'*). Confident that rural Russia would again soar from "the plow to the heavens," the poet declared that all Russians are rooted in the provinces, in "these rivers, peasant huts, and forests, from these countryside towns." Moscow was "awaiting new Gagarins from the provinces . . . reared like a simple, blue-eyed miracle, preserving the province, like a talisman in one's soul!"[79]

Meanwhile, Gagarin worship was partly rooted in the well-noted feeling of nostalgia for the supposedly good old Soviet days, but not only in this. Local boosters in Gagarin's hometown had hatched grand plans to create massive tourist complexes based on Gagarin's feat, and who could blame them?[80] If post-Soviet Moscow ran on oil and gas, memory was the fuel that powered the city of Gagarin, although not so readily convertible as oil into cold hard cash. Local politicians joined those efforts, as did Gagarin's relatives, one of whom emulated Gagarin's own conquest of Moscow. Gagarin's daughter Elena was appointed General Director of the Kremlin complex of museums in 2001, becoming, like her grandmother, a keeper of the nation's holy relics. Her connections to the heart of Russian power—along with the tireless efforts of cosmonauts to promote and protect Gagarin's image—have slowly drawn the center back into the province.[81] Indeed, if the province in the 1990s had scrupulously maintained the sacred national relics, ready to hand them back over to the center, the center in the twenty-first century seemed ready, at long last, to receive the gift. On July 31, 2008, Dmitrii Medvedev visited the city of Gagarin and laid flowers at one of Gagarin's many busts. He vowed that Moscow would celebrate the fiftieth anniversary of Gagarin's flight in grand style. He named 2011 "the year of Russian cosmonautics" and formed a commission (headed by Prime Minister Vladimir Putin) to develop a nationwide schedule of celebrations. The model of Gagarin's Vostok rocket in front of the former Cosmonauts' Pavilion of the former Exhibit of People's Economic Achievements was given a fresh coat of paint, cleansed of all graffiti, and unveiled to a national television audience on November 25, 2010, just in time for the fiftieth anniversary festivities. The national celebrations on April 12, 2011, suggested that Moscow was finally catching up to the patriotic spirit of the provinces—a national rebirth, uniting center and province, if only for a day.[82] Now if only every day could be Cosmonautics Day!

Notes

1. This article is adapted in part from my book *The Cosmonaut Who Couldn't Stop Smiling: The Life and Legend of Yuri Gagarin* (DeKalb: Northern Illinois University Press, 2012) and also uses excerpts, with permission, from my essay "Conquering Space: The Cult of Yuri Gagarin," in *Soviet and Post-Soviet Identities*, ed. Catriona Kelly and Mark Bassin (Cambridge: Cambridge University Press, 2012), 129–149.

2. "Dostizheniia inorodnogo khoziaistva. VVTs prevratilsia v iarmarku khlama," *Argumenty i fakty*, June 17, 2009, http://www.aif.ru/money/article/27498 (accessed June 18, 2009); interviews conducted by the author in Moscow in front of the former Cosmonaut Pavilion, June 23, 2007. The graffiti-covered plane was unceremoniously removed in 2008, causing a group of patriotic Russians to lobby on behalf of preserving the remaining rocket and officially declaring it a cultural monument. "Raketa 'Vostok' ne uletit s VVTs," *Izvestiia*, March 2, 2009, http://www.izvestia.ru/news/346000 (accessed April 18, 2009).

3. "Fake Gagarin," *Moscow Times*, October 5, 2007, http://www.themoscowtimes.com/arts_n_ideas/article/fake-gagarin/363249.html (accessed October 9, 2007). A lawyer for Gagarin's daughters, who sued to have all reference to their father removed from the movie, said, "Asserting that Yuri Gagarin entered into irregular sexual relations smears his name." The Gagarin daughters won the case.

4. Tatiana Khimikina (school teacher and museum guide in the Engels Gagarin exhibit), interview by the author, Engels, Russia, July 6, 2007.

5. "Iz istorii nauchnykh Gagarinskikh chtenii," *Gzhatskii vestnik*, March 9, 1999, 2; "Gagarinskim chteniiam—30 let," *Gzhatskii vestnik*, March 7, 2007, 3; "12-aprelia—Den' kosmonavtiki," *Gzhatskii vestnik*, April 11, 2006, 1.

6. S. N. Revin, "Obraz kosmonavta v predstavlenii starsheklassnikov," in *Gagarinskii sbornik: Materialy XXXIII obshchestvenno-nauchnykh chtenii posviashchennykh pamiati Iu. A. Gagarina 2006 g*, (Gagarin, Russia: 2007), 395; "Gagarin i Oskverniteli," *Slavianskii mir v Saratove*, February 22, 2007, 5; "Vesti Saratov," Evening Saratov Regional Television News, July 7, 2007, http://www.youtube.com/watch?v=Z2jhdenmcD8 (accessed February 14, 2013). On Gagarin's treatment in Russian culture during the 1990s, see Alexei Yurchak, "Gagarin and the Rave Kids: Transforming Power, Identity and Aesthetics," in *Consuming Russia: Popular Culture, Sex and Society since Gorbachev*, ed. Adele Marie Barker (Durham, NC: Duke University Press, 1999), 76–110.

7. "V pamiati—navsegda," *Gzhatskii vestnik*, March 9, 1999, 1; "Tekst ekskursii po vystavku 'Iz semeinogo al'boma'" (Y. A. Gagarin Unified Memorial Museum, Gagarin, Russia, 2004); *Gagarintsy o*

Gagarine (Gagarin, Russia: Polimir, 2006), 7, 10; "Put' v kosmos lezhal cherez Kalugu i Saratov," *Saratovskaia panorama*, March 10–16, 2004, 7; *Gzhatsk-Gagarin. 300 Let: Stikhi, poemy, pesni* (Moscow: Veche, 2007), 103.

8. "Moskva ot slez otvykla," *Kommersant*, November 23, 2007, http://www.kommersant.ru/doc.aspx?DocsID=828614&print=true (accessed November 26, 2007).

9. Rossiiskii gosudarstvennyi arkhiv ekonomiki (RGAE), f. 9453, op. 1, d. 35, ll. 164, 190, 192; *La Poeme de l'extase, Op. 54* by Alexander Scriabin, New York Philharmonic (2008).

10. RGAE, f. 9453, o. 1, d. 36, l. 53; o. 1, d. 37, l. 85; "12 aprelia 1961 goda ves sovetskii narod byl chastliv," *Nezavisimaia gazeta*, April 12, 2001, http://www.ng.ru/printed/16933 (accessed April 11, 2009); as cited in Vladislav Zubok, *Zhivago's Children: The Last Russian Intelligentsia* (Cambridge, MA: Harvard University Press, 2009), 132.

11. Paul R. Josephson, *Would Trotksy Wear a Bluetooth: Technological Utopianism under Socialism, 1917–1989* (Baltimore: Johns Hopkins University Press, 2010).

12. On the Russian and Soviet fascination with space flight before World War II, see James T. Andrews, "Storming the Stratosphere: Space Exploration, Soviet Culture, and the Arts from Lenin to Khrushchev's Times," *Russian History* 36 (2009): 77–87.

13. Michael Hagemeister, "Russian Cosmism in the 1920s and Today," in *The Occult in Russian and Soviet Culture*, ed. Bernice Glatzer Rosenthal (Ithaca, NY: Cornell University Press, 1997), 188.

14. James T. Andrews, *Red Cosmos: K. E. Tsiolkovskii, Grandfather of Soviet Rocketry* (College Station: Texas A & M University Press, 2009), 60–61.

15. Vladimir Makarov, interview by the author, Moscow, July 25, 2007.

16. RGAE, f. 9453, o. 1, d. 35, ll. 17, 67, 69, 205.

17. V. Ponomareva, *Zhenskoe litso kosmosa* (Moscow: Gelios, 2002), 243; RGAE, f. 9453, op. 1, d. 35, ll. 46, 166.

18. RGAE, f. 9453, o. 2, d. 21, ll. 13, 23, 28; d. 34, l. 25; op. 1, d. 35, ll. 232, 253.

19. RGAE, f. 9453, o. 1, d. 35, ll. 91, 151, 179; d. 36, ll. 17, 48, 55, 60, 66; d. 37, l. 64; Arkhiv Rossiiskoi akademii nauki (ARAN), f. 1647, o. 1, d. 260, l. 38.

20. RGAE, f. 9453, o. 1, d. 38, l. 1.

21. Asif Siddiqi, *The Red Rocket's Glare: Spaceflight and the Soviet Imagination, 1857–1957* (Cambridge: Cambridge University Press, 2010), 78.

22. Henri Lefebvre, *The Production of Space*, trans. Donald Nicholson-Smith (London: Blackwell, 1991), 98.

23. Virgiliu Pop, "Viewpoint: Space and Religion in Russia: Cosmonaut Worship to Orthodox Revival," *Astropolitics* 7 (May 2009): 150–63; RGAE, f. 9453, o. 2, d. 21, l. 31; d. 34, ll. 6–7; op. 1, d. 35, ll. 232, 236, 243.

24. RGAE, f. 9453, o. 2, d. 21, l. 31; d. 34, ll. 6–7; op. 1, d. 35, l. 251.

25. *Stranitsy istorii, Pokrovsk-Engel's*, 4th ed. (March, 2004), 7.

26. ARAN, f. 1647, o. 1, d. 256, ll. 9–10; RGAE, f. 9453, op. 1, d. 35, l. 218 (a poem from a letter writer entitled "A Song about a Russian Man").

27. RGAE, f. 9453, o. 1, d. 36. l. 62; o. 2, d. 21, l. 4; d. 34, ll. 9, 21, 138, 145, 175; d. 35, ll. 35, 215.

28. RGAE, f. 9453, o. 2, d. 34, ll. 5, 138; op. 1, d. 35, l. 78.

29. J. Nicholas Entrikin, *The Betweenness of Place: Towards a Geography of Modernity* (Baltimore: Johns Hopkins University Press, 1991), 65.

30. *Gzhatsk-Gagarin*, 54–56.

31. On this tendency, see Andrew L. Jenks, *Russia in a Box: Art and Identity in an Age of Revolution* (DeKalb: Northern Illinois University Press, 2005).

32. "Ot gurchalki do stantsii 'Mir,'" *Nezavisimaia gazeta*, June 9, 2001, http://www.ng.ru/printed/18481 (accessed April 11, 2009).

33. M. V. Stepanova, "Memorial'nyi muzei Iu. A. Gagarina," in *Gagarinskii sbornik: Materialy XXIX obshchestvenno-nauchnykh chtenii posviashchennykh pamiati Iu. A. Gagarina 2002 g*, part 1 (Gagarin, Russia: 2003), 244; Tamara Filatova (Gagarin's niece), interview by the author, Gagarin, Russia, August 8, 2007; A. V. Kniaginin, "Uchastie shkol'nikov i molodezhi v sobranii i izuchenii igr Iu. A. Gagarina," in *Gagarinskii sbornik: Materialy XXVII obshchestvenno-nauchnykh chtenii posviashchennykh pamiati Iu. A. Gagarina 2000 g*, part 2 (Gagarin, Russia: 2001), 96–97.

34. M. I. Gerasimova and A. G. Ivanov, eds., *Zvezdnyi put'* (Moscow: Izdatel'stvo politicheskoi literatury, 1986), 270.

35. Filatova, interview.

36. G. N. Mozgunova, *Syn zemli Smolenskoi* (Smolensk, Russia: Madzhenta, 2004), 25–26.

37. On such fears and the decline of the Russian countryside during and after the Soviet collapse, see Grigory Ioffe, Tatyana Nefedova, and Ilya Zaslavsky, *The End of the Peasantry: The Disintegration of Rural Russia* (Pittsburgh, PA: University of Pittsburgh Press, 2006).

38. *Gzhatsk-Gagarin*, 107, 134.

39. Mozgunova, *Syn zemli Smolenskoi*, 26

40. "Liubimaia sektsiia molodezhi," *Gzhatskii vestnik*, March 7, 2003, 4.

41. M. E. Kaulen, "Memorial'nyi muzei XXI veka," in *Gagarinskii sbornik: Materialy XXIX*, part 1, 226; M. V. Stepanova, "Memorial'nyi muzei Iu. A. Gagarina," in *Gagarinskii sbornik: Materialy XXIX*, part 1, 240–47; L. A. Filina, "Problemy memorial'nykh muzeev," in *Gagarinskii sbornik: Materialy XXIX*, part 1, 264.

42. Serguei Alex. Oushakine, *The Patriotism of Despair: Nation, War, and Loss in Russia* (Ithaca, NY: Cornell University Press, 2009).

43. "Tsenarii torzhestvennogo vechera, posviashchennogo A. T. Gagarinoi" (script for a ceremonial opening of an exhibit devoted to Gagarin's mother, Y. A. Gagarin Unified Memorial Museum, Gagarin, Russia, 2003).

44. Kaulen, "Memorial'nyi muzei XXI veka," 225.

45. Russell Schweickart letter, April 13, 1983, Association of Space Explorers, Folder 7, Hoover Institution Archives, Stanford, CA.

46. Filina, "Problemy memorial'nykh muzeev," 272; "Nizkii poklon etomy domu," *Gzhatskii vestnik*, March 9, 2004, 6.

47. Metodicheskaia razrabotka obzornoi ekskursii po domu-muzeiu roditelei Iu. A. Gagarina (guide for docents, Y. A. Gagarin Unified Memorial Museum, n.d.), 5.

48. "Rodnye," *Gzhatskii vestnik*, March 12, 1997, 2; "Zhizn', posviashchennaia pamiati kosmonavta No. 1," *Gzhatskii vestnik*, March 7, 2003, 5; Filatova, interview; T. D. Filatova, "Gagarinskaia gostinaia: traditsii Gagarinskoi sem'I," in *Gagarinskii sbornik: Materialy XXIX*, part 1, 274–77.

49. Filatova, "Gagarinskaia gostinaia," 275; "Nachalo nachal," *Gzhatsk-Gagarin*, no. 4 (April 2005): 1; "Novye vystavki," *Gzhatsk-Gagarin*, no. 7 (April 2004): 3.

50. "Mama," *Gzhatsk-Gagarin*, no. 3 (December 2003): 2; "Slovo o syne," *Gzhatsk-Gagarin*, no. 2 (May 2006): 3; Metodicheskaia razrabotka obzornoi ekskursii po domu-muzeiu roditelei Iu. A. Gagarina, 4; "Stsenarii torzhestvennogo vechera, posviashchennogo A. T. Gagarinoi," (script for the ceremonial opening of an exhibit devoted to Gagarin's mother, Y. A. Gagarin Unified Memorial Museum, Gagarin, Russia, 2003); Liudmila Demina, interview by the author, Gagarin, Russia, August 4, 2007.

51. Stepanova, "Memorial'nyi muzei Iu. A. Gagarina," 243.

52. "Grazhdanin vselennoi," *Gzhatskii vestnik*, March 9, 2004, 1.

53. Mikhail Gnedovskii, "Stsenarii ob'edinennogo memorial'nogo muzeia Iu. A. Gagarina" (Y. A. Gagarin Unified Memorial Museum, Gagarin, Russia, 1990), 3, 11.

54. Filatova, interview.

55. L. V. Leskov, "O geroicheskom entuziasme," in *Gagarinskii sbornik: Materialy XXV obshchestvenno-nauchnykh chtenii posviashchennykh pamiati Iu. A. Gagarina 1998 g* (Gagarin, Russia: 1999), 10, 14, 17; "Gorodskoi forum," *Gzhatsk-Gagarin*, no. 6 (March 2004): 1.

56. Ironically, it was funded in part by the billionaire financier and refugee from communism George Soros. M. V. Stepanova, "Iz opyta raboty . . . ," in *Gagarinskii sbornik: Materialy XXX obshchestvenno-nauchnykh chtenii posviashchennykh pamiati Iu. A. Gagarina 2003 g*, part 1 (Gagarin, Russia: 2004), 223.

57. "Gagarin. Kosmos. Vesna," *Gzhatsk-Gagarin*, no. 7 (March 2007), 3; "Istorii vystavki 'Gagarinskaia vesna,'" *Gzhatskii vestnik*, March 9, 2006, 3; M. V. Stepanova, "Nauchnaia kontseptsiia ekspositsii . . ." in *Gagarinskii sbornik: Materialy XXVII*, part 1, 51–52, 55.

58. Aleksandra Rossoshanskaia, interview by the author, Saratov, Russia, July 4, 2007; Khimikina, interview.

59. Stepanova, "Iz opyta raboty . . . ," 224–225. For Russian cosmist ideas and their influence on the presentation of Gagarin in memorial museums, see also *Turizm: problemy i perspektivy razvitiia* (Smolensk, Russia: 2006), 100–103: "Iu. A. Gagarin—znakovaia figura," *Gzhatsk-Gagarin*, no. 6 (March 2004): 1.

60. "Vystavka gzhatchan," *Gzhatskii vestnik*, March 7, 2003, 4.

61. *Gagarin-Gzhatsk*, 8; *Gorod Gagarin, Khudozhestvennaia galereia* (Moscow, 1990), 1; "Vystavka stanet mezhdunarodnoi," *Gzhatskii vestnik*, March 9, 2001, 3.

62. *Kosmicheskii aprel'* (Smolensk, Russia: Smiadyn', 2006), 126; "Iu. A. Gagarin: maloizvestnyi fakt biografii," *Gzhatsk-Gagarin*, no. 3 (March 2005): 3; *Gzhatsk-Gagarin*, 83–84.

63. "On vsekh naz pozval v kosmos," *Saratovskie vesti*, April 13, 2004, 1.

64. *Gzhatsk-Gagarin*, 91.

65. *Pravoslavnaia vera*, March 2004, No. 6, 1.

66. "A. Leonov: aktual'noe interv'iu," *Gzhatsk-Gagarin*, no. 3 (March 2005): 1.

67. V. A. Novikov, "Kakim viditsia v budushchem gorod pervogo kosmonavta," in *Gagarinskii sbornik: Materialy XXXI obshchestvenno-nauchnykh chtenii posviashchennykh pamiati Iu. A. Gagarina 2004 g*, part 2 (Gagarin, Russia: 2005), 327.

68. "Tak nash gorod otmechal iubelei Gagarina," *Gzhatskii vestnik*, March 5, 2005, 1; "Znamenitye gosti nashego goroda," *Gzhatskii vestnik*, March 9, 2005, 4.

69. "Gagarinskie chtenii v iubileinom godu," *Gzhatskii vestnik*, March 9, 2004, 2.

70. "Petr I na Gzhati," *Gzhatsk-Gagarin*, no. 12 (December 2005): 2; "Len," *Gzhatsk-Gagarin*, no. 13 (October 2004): 1; Novikov, "Kakim viditsia v budushchem gorod pervogo kosmonavta," 327, 340–41, 343; Mariia Stepanova, director of the Y. A. Gagarin Unified Memorial Museum, interview by the author, Gagarin, Russia, August 6, 2007.

71. "Turfir'my v gorode," *Gzhatsk-Gagarin*, no. 3 (March 2005): 1; "Obrashchenie k izbirateliam," *Gzhatsk-Gagarin*, no. 8 (August 2005): 1.

72. *Gorod Gagarin* (Smolensk, Russia: Smolensk Department of Culture, 2006); "Prirodnyi park 'Gagarinskii,'" *Gzhatsk-Gagarin*, no. 8 (August 2005): 1.

73. V. V. Bokhanov, "'Kosmoslend' . . . ," in *Gagarinskii sbornik: Materialy XXXIII*, 371–73; *Turizm: problemy i perspektivy razvitiia*, 81.

74. Bokhanov et al., "'Kosmoslend' . . . ," 377.

75. "Kurort imeni Gagarin," *Vedomosti*, March 7, 2008, http://www.vedomosti.ru/newspaper/print.shtml?2008/03/07/143181 (accessed March 10, 2007). With the passing of the fiftieth anniversary it is not clear if this capital infusion has occurred.

76. Yi-Fu Tuan, *Space and Place: The Perspective of Experience* (Minneapolis: University of Minnesota Press, 1977), 52.

77. Filatova, interview; "Gzhatsk, iun '61-go," *Gzhatsk-Gagarin*, no. 7 (April 2004): 2.

78. "9 marta—60 let so dnia rozhdeniia Iu. A. Gagarina," *Gzhatskii vestnik*, March 9, 1994, 1; "Obshchestvenno-nauchnye chteniia v iubileinyi den,'" *Gzhatskii vestnik*, March 16, 2004, 3; "Do planet—rukoi podat," *Saratovskie vesti*, April 12, 1994, 1; "K itogam XXXIV Gagarinskikh chtenii," *Gzhatskii vestnik*, March 30, 2007, 1; "Pervaia premiia imeni A. T. Gagarinoi," *Gzhatskii vestnik*, March 9, 2004, 6; "I budet vechnym ego podvig," *Gzhatskii vestnik*, April 6, 2004, 6.

79. "Obshchestvenno-nauchnye chteniia v iubileinyi den,'" *Gzhatskii vestnik*, March 16, 2004, 3; O. V. Prokhorenko, "Gorod Gagarin—vsemirno izvestnaia provintsiia," in *Gagarinskii sbornik: Materialy XXXIII*, 520–27.

80. "Vspomnim o nem," *Gzhatskii vestnik*, March 9, 1999, 4.

81. "Aleksei Arikhipovich Leonov i gagarintsy," *Gzhatsk-Gagarin*, no. 1 (March 2006), 2; "9 marta—73-ia godovshchina so dnia rozhdeniia Iu. A. Gagarina," *Gzhatskii vestnik*, March 9, 2007, 1.

82. "Iuru my nikogda ne zabudem," *Rossiiskaia gazeta*, April 10, 2009, http://www.rg.ru/2009/04/10/reg-roscentr/gagarin-chteniya.html, accessed April 11, 2009.

CHAPTER 5

Astronauts and Cosmonauts into Frenchmen: Understanding Space Travel through the Popular Weekly *Paris Match*

Guillaume de Syon

"A Historic Testimony You Will Wish to Keep" thundered the subhead of *Paris Match* (PM) in April 1961, shortly after Yuri Gagarin's successful orbital flight.[1] That statement summarizes best what a glossy weekly magazine could offer that other media could not: event coverage that was already processed into a historical memento. Although a marketing gimmick, the argument carried the day in a time when television sets were few and space travel was proving to be an intensely visual experience. Although the "space race" is mostly associated in memory with televised coverage, other media played a role in spreading the gospel of voyaging outward. For this purpose perhaps the best combination of visuals and words was the weekly glossy.

Life magazine's association with the National Aeronautics and Space Administration (NASA) astronauts comes to mind, but a similar phenomenon of bringing spaceflight to the living room table happened in France. Indeed, the first two decades of human space travel became facets of a global culture.[2] Considering their coverage in France helps us identify specific inflections in the production of signs for a localized readership.

This chapter examines the vision of human space travel that a widely read mainstream publication, *Paris Match*, offered the Francophone public between 1961 and 1981. *Paris Match* cast an essentially foreign event into one that was at home in France. In so doing it neither alienated nor polarized public opinion, yet, to borrow from Marshall McLuhan, it played a greater role in the transmission and understanding of a global visual culture of space into the French-speaking world than at any time since.[3] To understand the role of a weekly in this process, it is necessary to examine briefly its background and its role as a shaper of opinion.

Consuming *Paris Match*

Paris Match became a standard weekly magazine in the 1950s but occupied a unique niche in French culture that does not appear to have direct parallels in American media history. Often compared to *Life* magazine, which directly inspired it, *PM*'s format also incorporated elements seen in *Time* and the *Saturday Evening Post*, at least until the French magazine's acquisition by a new publisher in 1976. As such, the publication came to be viewed as a respectable form of entertainment in both working and middle-class households while stressing a moderately conservative agenda. Such a balance mattered in a patriarchal, culturally Catholic nation that was nonetheless opening slowly to the political left. The mix of glamour entertainment and news generally stressed human dimensions in all stories and pictures, and space travel was part of the mix. A key feature of the magazine involved the use of quadrichromy at a time when color photography remained very expensive in France. *Paris Match* relied partially on color illustrations to attract readers, especially through the careful selection of a cover photograph.

The weekly's cover usually sported local and world celebrities but also included human-interest stories of noncelebrities. As several historical and sociological analyses of *PM* have pointed out, the *PM* cover became a factor in the manufacturing of a modern French identity, whether it depicted such tragedies as the wars of decolonization or triumphs of athletes.[4] Until the 1980s, there were very few issues that would depict machines, but space exploration became part of the exception. Often, however, editors preferred to use images of "space travelers" to foster the acceptance of space travel.[5]

Selling up to 1.5 million copies per week in its heyday, *PM* served a unique function in the fifties, sixties, and seventies. It established itself as a purveyor of a constructed French identity, thus subscribing to a kind of social contract whereby its readers would receive information intelligible to all while enjoying the drama provided by photography. Furthermore, the era under consideration was not, in contrast to the American experience, a fully televised one. The first French presidential debates did not occur until 1965, and television channels remained state enterprises until 1982 (albeit liberalized in the 1970s). Furthermore, few households could afford a television set at the dawn of the "space age" in 1957, although the situation would evolve rapidly by the late 1960s. This gave printed media an edge in information and entertainment, with radio as the major competitor. The advertising style of *PM*, although targeting women at home in its most important pages (such as the back cover), also included publicity intended for men (notably fashion, cigarette, and car ads). *Paris Match* thus filled a niche that no other francophone weekly did (or chose to). It was socially acceptable to read it at the hairdresser but also fine to leave it in the home's receiving area. It became a mirror of idealized French identity, as semiotician Roland Barthes quickly noticed: in his seminal *Mythologies*, Barthes deconstructs everyday life and symbols of French culture by referencing the popular press, especially "Match," the moniker associated with *PM*.[6]

Barthes's analysis asserted that the magazine cautiously presented events so as not to upset readers and cast middle-class ideals as universal ones.[7] The point is convincing when it applies to ideological discussions of specific French events, but France did not become a space power until 1965, and NASA has always been better known there than Centre National d'Etudes Spatiales (CNES), its French counterpart. An alternative tool, Pierre Nora's notion of "dramatic entertainment" to justify the existence of grand technological projects, would likely fit the bill of France's nuclear agenda or even specific machines like the Concorde.[8] But this does not apply as well when foreign technical endeavors fascinate an international audience. The human exploration of space expresses one of many historical forces of modernization and contributes to the transnational process of globalization—I borrow here from sociologist Anthony Giddens. He notes that in any human system, however, the process is imperfect and depends heavily on such transmission factors as the media. Giddens suggests notably that, within globalization, local happenings are shaped by events occurring many miles away and vice versa. As Giddens further notes, however, this practice is neither homogeneous, continuous, nor necessarily successful and depends on which systems of knowledge are exposed to this cross-culturation.[9] This dialectical process with uncertain outcomes contributes to the theme of transnational understanding of human space exploration.[10]

Bringing Space to France

Like many media in the 1950s, *PM* covered matters related to space selectively, with an emphasis on a mix of entertainment and education. The 1953 Fourth International Astronautical Congress in Zurich, for example, offered ideal fodder to discuss space travel. Taking advantage of the summer lull (the congress took place in early August), *PM* published a multipage color "filler" article that sought to summarize the scientific discussions but emphasized, of course, the most fantastic one: by 1980, humanity would reach the moon: "The first man to walk there has already been born."[11] The coverage was remarkable on three grounds. First, France was ending its bloody war in Indochina and about to embark on another one in Algeria, all the while experiencing political instability. Discussing space was not on the public's mind, so this was at best sophisticated entertainment. Second, there was no French delegation present, as Alexandre Ananoff's Groupement Astronautique Français, a founding organization of the congress and of the International Astronautical Federation, had just disbanded.[12] Finally, none of the papers read at the conference discussed a trip to the moon. Rather, as *PM* columnists acknowledged in an introductory aside, the suggestion of a lunar trip was the result of cautious optimism based on postpanel discussions.[13] Yet this was a way to filter a tangible scientific discussion while offering entertainment. Columnist Raymond Cartier would capitalize on this as the cold war in space began.

Two years later, the Sixth International Astronautical Congress met in Copenhagen and became a proverbial stepping stone for a widening of public awareness about space. Raymond Cartier picked up on the sensational announcement on July 29, 1955, by President Eisenhower that the United States would start work on orbiting a satellite.[14] What would have been a quiet scientific gathering became a media circus. Casting a critical eye on the excitement that accompanied the proceedings, Raymond Cartier nonetheless admired the "Disney line" as he chose to call the series of machines presented in *Collier's* magazine and on the Disney television show. A wider slice of the French public thus experienced wonderful illustrations by artist Chesley Bonestell.[15] The timing of the article, however, in mid-August when much of France shuts down to sun itself meant that not as many people read it. Still, such coverage in a mainstream publication made notions of human space travel as a "world of tomorrow" increasingly normal.[16]

Although the very act of space travel was phenomenal enough to warrant reporting, *PM* coverage emphasized time and again the human element. The magazine's covers maximized the human face, and when machines were depicted, they usually constituted an exception on the basis of uniqueness or the perceived historical nature of the event they were associated with. Pictures of a new Citroën design in the 1950s and of an American aircraft carrier during the Cuban Missile Crisis offer cases in point, as do such covers as Concorde's first flight.[17] With regard to spaceflight, the same pattern was followed. Sputnik's orbiting warranted an artistic commissioning of a reddish metal ball on the cover, but the Earth in the background implied an unseen human element.[18] The articles accompanying the event coverage, however, relied on photographs of astronomer Alexandre Ananoff explaining orbits with chalk and board illustrations: the reader was back on a school bench. Perhaps more entertaining, an accompanying piece discussed the establishment of a manned laboratory on the moon, thus displacing the robotic aspect of the Soviet achievement. Humans may have stood behind machines, but their presence seemed to be required to understand their function.

Taking France Out of Space

Paris Match's emphasis on people meant that the human space race would be a perfect way of making it and its associated technology understandable to lay readers. However, where French space efforts were concerned, there was little that could be done. The creation of CNES, France's space agency, in 1962 made the daily news but not *PM*. Having Minister Pierre Guillaumat, one of the fathers of the French nuclear program, wax poetic about the likes of Channel flyer Louis Blériot exploring space may have echoed President de Gaulle's grandiose visions of France, but the timing was off.[19] All that remained was the trope of stressing that France would be the third power in space (notwithstanding the fact that other satellites had flown aboard American rockets).

Covers, for example, were problematic. Always careful to emphasize the human element, *PM* considered rocketry newsworthy, but the need to keep its target audience as wide as possible may have influenced the very limited display of missiles. This sometimes led to comical choices. The Véronique booster on its pad in the desert became the main cover photograph, but two insets showing a would-be princess and a fashion model balanced out the decidedly technical shot of a missile with a female name.[20] Comparison, when it happened, was done in geographical terms to explain the importance of missile and satellite research. In the same issue, the testing range at Hammaguir in the North Africa desert thus became a "Saharan French Canaveral."[21] Even when the French space program lofted its first satellite in 1965 in the midst of firsts by the American Gemini and Soviet Voskhod human spacecraft, the article explaining the birth of the "third space power" chose an old trope: present the satellite's designers. A similar approach had existed in the late 1950s to show France's nuclear efforts.[22] The satellite's AS-1 dry acronym had already evolved into the Astérix moniker from a popular comic strip satirizing France through its Gallic past. Pictures of the rocket launch had their place in the magazine, naturally, but the satellite's "four buddies" article on the designers ran fourteen pages compared to the launch's four.[23] Overall, then, although space achievements were acknowledged, the fact that France had not considered entering the human space race meant that readers of *PM* would experience a limited account of French space efforts while the superpowers enjoyed full coverage. The angle *PM* chose, however, would acknowledge the space race while deemphasizing the cold war.

Paris Match's format emphasized photography, but it had dedicated columnists specializing in human-interest stories. When it came to "global" events, several names became associated with the space program. Raymond Cartier, who presented the cold war in all its facets to readers, also became a kind of op-ed writer on space, although he was often seconded by reporters Paul Mathias, Marc Heimer, and Jean Prouvost. Their writing supplemented the syndicated stories acquired from *Life*, notably astronauts' testimonies. In doing so, these writers also introduced new forms of expression into the French language as they sought to offer understandable contexts to the astronauts' experiences.

Humanizing the wonders of space travel was an added challenge for all non-English or and non-Russian media. Translating these into French became almost comical. As postwar France had experienced accelerating foreign exposure while failing to keep up with Shakespeare's tongue, readers expected a precise translation into French to the exclusion of any Anglicism. For the leader of "Project Mona" (a codename given by the reporter to a project that was referred to officially as Pioneer zero), an attempt to hit the moon in August 1958 was not conducting a test, but an orchestra. His workplace, Cape Canaveral, was likened to a "train station" dispatching machines onto orbital "tracks" known as "rails" (the only term available in French).[24] Some words defied any comparison with ground transportation, leading to the a new verb, desatellize (désatelliser). Overall, writers creatively contorted words that would not warrant translation now but required clarification then.[25] Paradoxically, columnist Marc Heimer's attempt to bring technical terms to the lay public meant that he became the first to translate some of these terms, such as aluminized nylon to described spacesuit fabrics.[26] The experiences described were so novel that some explanations clearly caused confusion for their authors. John Glenn's "we are go" was carefully explained as astronaut slang for saying everything was in order, and his capsule's fiery reentry was likened to a skidding automobile.[27] On the other hand, countdown (referring to Gagarin's launch) was one of the few words left in English, although a proper translation would soon exist.[28] All these metaphors helped bring the space experience to the readers, but picturing the humans associated with space was clearly key to accepting this new dimension of human evolution.

Enter the Rocketmen

Early coverage of space efforts did include picturing engineers. The coverage of the Pioneer project presented several men associated with the program. Their status as fathers was what interested the editors. A Captain Griffith described as in charge of launch operations was pictured with his five children, who all were convinced that any lofting missile was "daddy's rocket." Meanwhile, Theodor Gordon, an engineer who played a key role in designing the Thor booster, was shown biking with his two kids. Although they represented patriarchy, it was not linked to masculine technological empowerment. The commentators likened them to "pères tranquilles" (easygoing dads) who went to work on rockets instead of on assembly lines.[29] Much as Griffith and Gordon may have come across as decent fellows, they were anonymous and thus unlikely to embody anything past the photographs that accompanied the failure of their rocket to reach orbit.

Although French enthusiasts of the postwar years remember the efforts of astronomer Alexandre Ananoff in presenting space, the best known "spaceman" in France at the time, and the charismatic figure who would most dominate *PM* space coverage before the emergence of the astronauts and cosmonauts, was a German-American. Wernher von Braun likely had the peculiar distinction of being the most famous German in France besides Chancellor Adenauer, and his savvy media image definitely played a role in the extensive coverage *PM* gave him. Still, the abundance of interviews and pictures of him, although associated with the American space program, upset some readers. In a country where the 1962 Franco-German treaty of friendship was mostly ink on paper aside from student exchanges, generations of Frenchmen had difficult memories of the world wars and their neighbors. Von Braun's notoriety was known early on. In 1953 he had been described as "Dr. V2," but the narrative that accompanied the moniker emphasized the myth of the scientist duping Hitler.[30] The successful orbiting of the first American satellite, Explorer 1, on February 1, 1958 (French time), gave rise two weeks later to a presentation in *PM* of the "inventor of the V-2."[31] In fact, this was only the latest iteration of von Braun's past work. Subsequent coverage continued to be positive and included "the prodigal child's" swift trip to Bavaria to visit his parents.[32] Because the article, like later ones, emphasized the fascination of space travel, it was easier to put on a positive spin that downplayed von Braun's past. In fact, he was *PM*'s go-to expert, and he either wrote articles, answered queries, or was featured on at least six more occasions between 1959 and 1965.[33]

The January 9, 1965, interview, however, and subsequent mentions that year prompted some readers to react negatively. The magazine ran a series of articles commemorating the end of World War II, and for the first time, it mentioned death and concentration camps. It is not clear whether this was the tipping point, but survivors of the Dora slave labor camp where the V-2s were assembled wrote the magazine to protest the praise it was lavishing on von Braun. Survivor Yves Béon remembered thirty years later how he had very little success pointing out the issue: "Someone at the paper received me and explained: 'You are nobody. Von Braun is really somebody. If we say anything against him, US firms will stop advertising. Sorry. Goodbye.'"[34] The fact that in 1966 *PM* underwrote the translation and publication of von Braun's book on *The History of Space Travel* tends to confirm Béon's recollection. So does the fact that instead of investigating the charges of the survivors, the editorial board offered von Braun an opportunity to respond to his detractors and accepted his assertion that he was appalled by, but not responsible for, the atrocities. Nothing about the matter was published.[35] Fortunately for von Braun, fascination with space travel had shifted to astronauts and cosmonauts, and the challenges to his troubled past did not return before his death, at least not in the pages of *PM*: Coverage of his persona remained positive in all subsequent issues, although his name no longer appeared directly on the cover of the magazine until his death in summer 1977.[36] Although he remains to this day a household name, stardom did not become associated with him alone, but also with some of the people he helped get into space.

The Space Traveler Imperative

The cross-cultural assimilation of space travel required casting astronauts and cosmonauts as stars. As John Gaffen and Diana Holmes have noted, the concept of stardom in postwar France had resulted in some peculiar choices of heroes to admire. Simply put, alongside the usual actors and singers stood bicycle racers, intellectuals, a movie maker, and President Charles de Gaulle. They were the expression of a nation whose World War II humiliation would not be discussed until the 1980s. The search for symbols in the face of a rapid capitalist growth that accelerated secularization may help explain the diverse lot French popular culture embraced in the 1960s.[37] A few Americans did make the cut, such as President Kennedy, whose bon mot about simply being with Mrs. Kennedy on a visit to Paris made him temporarily more popular than Charles De Gaulle. But love of United States was complicated in the 1960s, partly because of the threatened sense of identity some French experienced.[38] Neither could the Soviet Union expect much sympathy from a strongly anti-Communist middle class. Astronauts and cosmonauts, however, became the proverbial exception to any misgivings. Their very place of work made them into stars, but the angle of coverage *PM* chose to mediate their activities sought to demystify the heroic dimension.

With astronauts and cosmonauts becoming proverbial stars, coverage of their prowess was a logical step. To ensure maximum coverage, *PM* did sign licensing agreements with American press syndicates and time and again informed its readers that alongside the daily *Le Figaro*, it had exclusive access to the feats of astronauts. This "achievement" was even announced on the front cover of a June 1965 issue.[39] Such a deal, which essentially came down to purchasing coverage from press magnate Henry Luce's Time-Life conglomerate, did indeed allow a unique diffusion of astronaut stardom into France.

Seeing astronauts and cosmonauts and understanding their behavior is what brought space home to France. *Life* Magazine's coverage of the astronauts and their families, when syndicated and translated, gave readers plenty of information and in many ways reassured them. *Paris Match* supplemented these shots with those obtained from the Soviet news agencies. Stripped of their spacesuits and far from the launch pad, astronauts and cosmonauts looked very much like Frenchmen. The very act of making them *look* normal actually contributed to both their greatness and to their acceptance. This was the case with the seven Mercury astronauts who would first try to beat the Soviet Union into orbit.

They failed. Yuri Gagarin was first with his flight on April 12, 1961. Young and with a winning smile, he gave off an ideal image of "homo sovieticus" rising from the ruins of Stalinism. Offering insights into the man, however, proved difficult. Despite the limited information on the Soviet space program, *PM* editors made valiant efforts at balancing the coverage of both American and Soviet programs. Although Yuri Gagarin rated the front cover for obvious reasons, a second cosmonaut, Gherman Titov, who orbited for an entire day in August, also made the front page even though the Berlin Wall had just gone up.[40] The editors had recognized that regardless of ideology, the achievement required acknowledgement. *Paris Match* columnist Marc Heimer had to exercise much patience in obtaining authorization to interview them. A sighting of Yuri Gagarin dining in Paris had resulted in little more than a paparazzi-style photograph and a very plain interview, where the hounded cosmonaut explained that he and his colleagues were but ordinary men.[41] Although the statement was true and even endearing, the interviewer noted with frustration that the presence of "minders" may have had as much of an impact on Gagarin's awkwardness as the spotlight did.[42] *Paris Match* would have better luck covering the success of Alexei Leonov.

FIGURE 1. Leonov leafs through the *Paris Match* issue that bears his picture (issue 833, March 27, 1965). The photograph was taken in Moscow during an interview granted to Marc Heimer, the magazine's science correspondent. Photo André Lefebvre PM 835 © Hachette Filipacci Associés for Paris-Match, reprinted by permission.

Leonov's space walk in March 1965, the first ever, earned him front page coverage, and both luck and a subsequent interview gave him fame on the scale of Gagarin's in France. Within ten days of his flight, Leonov's portrait graced the *PM* cover (Figure 1). The crispness of the image and its good contrasts that framed a smiling, focused face peering out of the pressure suit drew reader reactions. Other cosmonaut shots had a slightly opaque feel to them, characteristic of Soviet press photography. How, a reader's letter wondered, had the staff obtained such a great shot on such short notice? It was, *PM* admitted, sheer luck that had allowed them to obtain a better quality picture.

That picture may have contributed to the granting of an interview with Leonov. At the meeting, the cosmonaut received a copy of the issue. Another photograph was taken of him leafing through the magazine, and it was duly published, as were his broad answers to the interviewer's precise questions. Journalists covering Soviet aerospace program have joked for decades about the veil of Soviet secrecy that delivered such nuggets of information as a space capsule's altitude reaching "very high." Asif Siddiqi has recently clarified how difficult such restrictions were on the cosmonauts, who had become objects of admiration while being forced to keep silent.[43] Leonov, who in earlier coverage had been shown sketching and exclaiming he wanted to paint while in space, offered a novel answer to his interviewer. When the latter asked him to describe the tethering to the capsule, Leonov took the fountain pen handed to him and sketched himself hanging outside the capsule by an ink thread. The doodle became the embodiment of the interview.[44] Although no technical information had been handed over, the act of drawing brought about smiles and registered well with readers (Figure 2).

The stress on the human element to the point of playfulness diffused the frustration *PM* reporters experienced with the Soviet space program. The limited enthusiasm was palpable in the plain coverage Valentina Tereshkova's 1963 exploit elicited. Secretly launched, like her male comrades, her story was

FIGURE 2. In response to Marc Heimer's query about his position before the onboard camera, Leonov took the journalist's pen and began sketching himself, his spacecraft, and the angle of the sunlight (lower right). He dedicated the finished drawing to the *Paris Match* readership. Photo André Lefebvre PM 835 © Hachette Filipacci Associés for Paris-Match, reprinted by permission.

filtered through the *Tass* and *Novosti* press agencies. The first woman in space was presented in but a few pictures. Her marriage to another cosmonaut the same year she flew, however, made the pages of the French magazine in full color. What could have been an element of inspiration for women reading *PM* at the dawn of the new feminism was reduced instead to a further affirmation of patriarchal tradition, in contradiction to the claimed Soviet message of gender equality. Patriarchy also permeated coverage of the American space program.

As the first American to enter space, Alan Shepard became a household name in France too. The picture of him walking to the booster was, of course, cover material, but *PM* quickly capitalized on the *Life* approach with a full spread about Louise Shepard, the astronaut's wife, meeting with Jackie Kennedy. The first lady was a household name in France, and the interaction of the women turned into one of the article's foci. The other was the Shepard family praying around the dinner table.[45] Although the astronaut's religion was not divulged, the Christian stance evoked a traditional French household in the countryside and likely contributed to further endearing the astronaut in French eyes. Other Mercury astronauts were the focus of similar selections of photographs. Scott Carpenter kissing his wife before going to work (launch) offered a further reinforcement of tradition. The men would also toast each other before a flight (although French readers were not told the drink was grape juice). All these images showed masculinity that was strong, but not overbearing: later coverage of Gemini astronauts showed them as working in space but going home on weekends to be dads.[46] In another case, the clergyman dining at an astronaut's request even reinforced further traditional visions of patriarchy: Even as France's churches were fast emptying, the cultural resonance of a table prayer photograph was too special to pass up and further suggested the men flying into space were no different than those driving to an office.

After the successful completion of the Mercury missions, there was a similar fascination with Project Gemini, a key stage to preparing for the Moon landing by placing two men in orbit for several days. The first Gemini crew became twins indeed in *PM*'s coverage through descriptions of their weekend prior to launching.[47] Still, the increased pace of space firsts did push aside some of the human-interest stories in favor of watching the race unfold. Although the successful spacewalk by Ed White two and a half months after Leonov occasioned a humorous "I can walk in space, too" title, Raymond Cartier went about summarizing the stakes in the space race.[48]

The year 1965 marks not only the last year of important firsts for the Soviet human program and the beginning of the American rise to dominance but also the start of a more critical appraisal of Soviet and American efforts in *PM*. Accompanying a kind of family chart of who's who in space travel, Cartier's analysis cast squarely the advantages and shortcomings of each program and now raised questions about their values. Much of what Cartier said, including the issue about Soviet space operations being a military secret, was not new, but it marked a more public stance in *PM*'s supposedly apolitical coverage of space-related endeavors. The Soviet wall of secrecy had tried the patience of reporters, and several, including Cartier, voiced their approval of American openness. Commenting on the element of spectatorship as well as military prowess, Cartier, who had served for years as the cold war watcher for *PM* readers, placed new emphasis on the fact that the race was more than a sport.[49] In so doing, he openly expounded upon the trope that there existed a knowledge imperative to reach the Moon. Otherwise known as the "science dividend," the claim offered a way to affirm an ideal and avert criticism. In doing so, Cartier followed the established line that semiotician Roland Barthes had criticized years earlier, whereby event coverage was always affirmative to the point of wonderment, as a means to avoid shocking readers.[50] Perhaps the best example of such "affirmativeness" is in the coverage of the Apollo missions.

Apollo's Appeal

Each of the manned Apollo missions enjoyed special coverage in *PM* and affirmed the adventure of exploration. The end of Soviet supremacy in 1966 had not gone unnoticed as Gemini missions drew to a close.[51] Space coverage would be de facto American, except when tragedy struck the cosmonaut corps.[52] Apollo 8, the first mission to circle the Moon, received as many front covers as Apollo 11, whereas the preparatory missions Apollo 9 and 10 each got two. In the case of Apollo 8, aside from "Earthrise," one of the most famous pictures of the twentieth century, the last issue devoted to the mission displayed mission commander Frank Borman in an unlikely pose. Visiting Paris in early 1969, Borman was photographed standing, answering questions with his arm raised as if to protect himself from the camera glare. The title the editors chose was "these eyes had seen the Moon."[53] The messiah-like tone of the title page echoed perhaps Borman's own reading of Genesis verses on Christmas Eve in lunar orbit, but the choice of imagery was fascinating too. Was it the commander in a dark suit looking upward that captured the editors' choice? Bormann and his crewmates had gone to another world. Stressing the human dimension as the last vision of the mission was what counted most.

Although Apollo 9 and 10 have been cast as "dress rehearsal" operations, they became part of the *PM* ritual of consecrating the astronauts who flew them. Apollo 11 naturally warranted dozens of pages and several covers, but although it was deemed the ultimate exploit, Apollo 12 symbolized the confirmation of the triumph. To keep reader interest high, *PM* decided to commission the cartoonist Hergé to draw a four-page summary of the mission. *Tintin*'s creator thus embarked on the only nonfiction work of his career, depicting lightning striking the rising Saturn V rocket and also showing pictures that could not be photographed yet, such as the launching of the lunar module's upper stage from its lower stage. To balance what some adult readers might have called child's play (even though drawn by a respected master of the comic arts), *PM* also had a portrait of the Apollo 12 crew appear on the front cover. The effort comes across as forced, as if to balance out the five covers that accompanied Apollo 11, but it still reflects the acknowledgment that space travel was a unique endeavor. At a time when European youth were protesting U.S. actions in Vietnam, NASA's achievements were among the few things generally ignored by protesters. By emphasizing such success in each of the subsequent Apollo missions (including the "successful failure" of Apollo 13), *PM* affirmed a tacit support for some American values. This attitude was exacerbated by Soviet secrecy.

Space tragedy also brought coverage. The Apollo 1 fire that killed three astronauts in 1967 yielded the expected article from the specially dispatched reporter in a tone of respect and sadness similar to that found in *Life* magazine.[54] The deaths of cosmonauts also became the subject of speculation because of secrecy issues, but the content varied in tone. Gagarin's plane crash in 1968 was described as "stunning" on the cover, but the substance of coverage bore nothing particularly out of the ordinary: speculation and questions were all the report could offer in view of Soviet reticence.[55] On the other hand, Vladimir Komarov's tragic end on the first test flight of a Soyuz a year earlier and that of three Soyuz 11 cosmonauts returning from the first space station, Salyut 1, in 1971 warranted unique coverage. Marc Heimer had met Komarov after the latter's first flight in 1964. In an obituary of sorts, Heimer stressed his personal admiration for the dead cosmonaut and his dislike of the secrecy that surrounded Komarov's crash; the harshness of the comment suggested that Komarov's friends deserved the truth.[56] This was the closest *PM* had come to breaking its political neutrality when covering space matters. The death of the Salyut 1 crew also made the cover of the magazine at a time when interest in space was waning, and there was no follow up. The event was just that, and no significance was assigned to it beyond the crewmen's smiling faces on the cover.[57]

The end of the Moon race and the ongoing social tensions of the post-1968 student rebellion also coincided with a sharp drop in sales of *PM*. New media outlets, changes in editorial practices, and expanding television coverage accelerated the magazine's decline as well as shifting reader interest. Similar forces produced the rapid decline of *Life* and other large-format picture magazines in the United States. Space topics remained newsworthy and included the announcement of the shuttle development, the orbiting of the Skylab space station, and the Apollo-Soyuz Test Project. However, although Skylab made the front cover and was described as a "house in space," the U.S.-Soviet rendezvous in space in 1975 was given only a small inset photo next to the big news of summer: Christina Onassis' engagement.[58]

This shift in reporting practices reflects the magazine's transfer to a new publisher in 1976 with a corresponding style change. Although continuing to emphasize the quality of photography that had branded its reputation, *PM* reduced the number of editorial and background articles in favor of more sensational and shallower reporting. When dramatic, space events continued to offer grounds for coverage, they were placed front and center, but rarely on the cover. Although the golden age of the shuttle era made spaceflight seem routine, coverage of space remained a boon for *PM*. It even exceeded the million copy mark of its 1960s heydays for some of its reports on the preparation of French "spationaut" Patrick Baudry for a shuttle flight in 1985.

The (French) Space Traveler Becomes a Spationaut

The ultimate irony of the perceived need to filter the space travel experience for a French public likely found its expression in the appearance of the word spationaut. Its first use is unclear, but *PM* and other French media (especially TV and radio news) were quick to adopt it from 1980 onward to designate the Frenchmen scheduled to go into space. Its creation was likely a direct result of the cold war. France's agreement with the Soviet Union to fly a Frenchman into orbit occurred in 1979 and preceded a similar agreement with NASA.[59] That the Soviet Union invaded Afghanistan soon after did not help matters for France, whose peculiar foreign policy stance made it a Western ally that claimed a "third way" in the cold war.[60] Although the very term astronaut had a francophone root, spationaut would become the standard term of reference. Ironically, it did not enter the official government glossary until 1995 and has yet to appear in the elite dictionary of the French Academy.[61] The official announcement of the selection of spationauts Jean-Loup Chrétien and Patrick Baudry also saw the term enter common use as the two joined cosmonauts at Star City and astronauts in Houston. Coverage of Chrétien's training, with Baudry as backup, drew interest, but Soviet secrecy limited exposure until Chrétien's flight to the Salyut 6 space station in 1982. On the other hand, because journalist Benoît Clair had been allowed to follow the spationauts over several months during their NASA training (Baudry eventually flew on STS-51G in 1985, with Chrétien as his backup), readers were treated to a multiple-installment documentary that proved so successful it was republished in book form.[62] Since then, however, *PM* has not covered space with the same degree of success. Other media have supplanted its once enviable position. Instead, the magazine now aims for a different kind of star, generally a singer, a sports personality, or a crowned head.

Conclusions

The first two decades of human space travel are part of the modernity of globalization in that they also help us understand the impact of print media in France in covering such events. Televised images, although important, remained limited in scope because of a state monopoly.[63] Glossy weeklies like *Paris Match* thus played a greater role in the transmission and understanding of a global visual culture of spaceflight and of associated technologies. The magazine's ideology of "humanness" made a seemingly impossible technology appear acceptable, perhaps even welcome. The emphasis on the human element associated with boosters thus mediated knowledge about the space race in a manner the nonhuman space adventure could not achieve. Paradoxically, by helping France acclimate to the global village of space travel, media such as *PM* may have mediated its impact, thus confirming Giddens's caution about the instability of such agencies. In fact, the intensity of the coverage may have achieved the effect Howard McCurdy observed in the United States: it made space travel seem routine. The heroes thus disappeared.

Few in France aside from space enthusiasts remember the names of the country's first spationauts. The term itself, although officially used at CNES, has lost its meaning as a national identifier. In his autobiography, Jean-Loup Chrétien purposely refers to himself as a "French cosmonaut."[64] Yet many remember earlier "space travelers" named Gagarin, Leonov, or Armstrong. By the same token, however, one could suggest that this was precisely the highest achievement of *Paris Match*'s coverage: turning astronauts and cosmonauts into Frenchmen facing challenges in their jobs, tending to their families, laughing, and going on: in other words, assuming the routine of everyday life. Many French readers thus came away envisioning space travel with an uncritical and welcoming eye that also helped pave the way for France's fielding of spationauts aboard Soviet and American spacecraft.

Notes

1. *Paris Match* (hereinafter *PM*), no. 628, April 22, 1961, cover.

2. Martin Collins, "Afterword: Community and Explanation in Space History (?)," in *Critical Issues in the History of Spaceflight*, ed. Stephen J. Dick and Roger Launius (Washington, DC: NASA, 2006), 603–13.

3. Marshall McLuhan, *Understanding Media: The Extensions of Man* (New York: McGraw-Hill, 1964), 27. Rémy Rieffel, "Les charactéristiques et la spécificité de la presse magazine en France," in *Die Zeitschrift—Medium der Moderne*, ed. Clemens Zimmermann and Manfred Schmeling, Frankreich-Forum 6 (Saarbrücken: Universität des Saarlandes, 2005), 47. Although Switzerland and

Belgium each had their own glossy weeklies, religious and linguistic divisions in both limited their sales, whereas *PM* was often available in major cities outside France.

4. See Nicola Cooper, "Heroes and Martyrs: The Changing Mythical Status of the French Army during the Indochinese War," in *France at War in the Twentieth Century: Propaganda, Myth, Metaphor*, ed. Valerie Holman and Debra Kelly (New York: Berghahn, 2000), 126–41; Philip Dine, *French Rugby Football: A Cultural History* (New York: Berg, 2011).

5. The term in the original French, "les voyageurs de l'espace," was first used in summer 1953 in a report on the Fourth International Astronautical Congress. Philippe Baleine, "La conquête du ciel. Les hommes feront leur voyage dans la lune en 1980, concluent les savants de l'atronautique réunis à Zurich," *PM*, no. 231, August 22, 1953, 26–27.

6. Roland Barthes, *Mythologies* (1957; repr., Paris: Seuil, 1970), 93–94, 105–6, 201–2.

7. Graham Allen, *Roland Barthes* (London: Routledge, 2003), 37.

8. Gabrielle Hecht, *The Radiance of France, Nuclear Power and National Identity after World War II* (Cambridge, MA: MIT Press, 1998), 13–14. Hecht's landmark study thoroughly documents the locale of the French nuclear program, but her choice of publications to reflect the national mood does not include the most mainstream weeklies in an attempt to move away from their "Parisian" flavor.

9. Anthony Giddens, *The Consequences of Modernity* (Stanford, CA: Stanford University Press, 1991), 7–8, 61–64.

10. David Gauntlett, *Media, Gender and Identity: An Introduction* (London: Routledge, 2002), 102–3.

11. Baleine, "La conquête du ciel," *PM*, no. 231, August 22, 1953, 24–30.

12. 4th International Astronautical Congress as remembered by F. Casal, Rapperswil-Jona, Switzerland: 19 October 2010. http://ftp.iafastro.com/index.html?title=IAC1953 (accessed January 28, 2012).

13. Baleine, "La conquête du ciel," *PM*, no. 231,August 22, 1953, 25.

14. David S. F. Portree, *NASA's Origins and the Dawn of the Space Age*, Monographs in Aerospace History 10 (Washington, DC: NASA, 1998), http://history.nasa.gov/monograph10/ (accessed January 9, 2012).

15. Bonestell, famous for his technical illustrations and, later, space paintings in *Collier's Magazine*, had seen some of his work published on the cover of the French translations of space writer Willy Ley and on the French edition of *Popular Mechanics* (*Mécanique populaire*). This was the first time, however, that his art appeared in wider circulation.

16. Raymond Cartier, "Le monde de demain commence aujourd'hui," *PM*, no. 334, August 20, 1955, 54–59.

17. "Blocus de Cuba, en couleurs à bord du porte-avions atomique Enterprise," *PM*, no. 708, November 3, 1962, cover.

18. Special Issue on Sputnik, *PM*, no. 445, October 19, 1957.

19. Claude Carlier and Marcel Gill, *Les trente premières années du CNES* (Paris: CNES/La documentation française, 1994), iii. See also Walter McDougall, "Space-Age Europe: Gaullism, Euro-Gaullism and the American Dilemma," *Technology and Culture* 26 (1985): 179–203.

20. "Au Sahara la fusée française," *PM*, no. 519, March 21, 1959, cover.

21. Jacques le Bailly, "Canaveral français au Sahara," *PM*, no. 519, March 21, 1959, 64.

22. Georges Menant, "La France atomique," *PM*, no. 500, November 8, 1958, 88–106; " Reggan, la ville atomique française," *PM*, no. 566, February 7, 1960, 46–49.

23. "Décollage d'Astérix de Hammaguir," *PM*, no. 869, December 4, 1965, 4–7; "Les quatre copains du satellite," *PM*, no. 870, December 11, 1965, 52–65.

24. *PM*, no. 489, August 23, 1958, cover. The name Project Mona that the French reporter used was, in fact, the designation given by the Advanced Research Project Agency (ARPA). The official name became Pioneer once NASA officially assumed control of the project shortly thereafter.

25. Louis Guibert, *Le vocabulaire de l'astronautique* (Rouen, France: Presses universitaires de Rouen, 1967), 84.

26. Guibert, *Le vocabulaire*, 78.

27. Guibert, *Le vocabulaire*, 114; Marc Heimer, "5-4-3-2-1-0 lâchez tout!" *PM*, no. 673, March 3, 1962, 47.

28. *PM*, no. 628, April 22, 1961, cover photograph of Cosmonaut Yuri Gagarin.

29. "Objectif lune." *PM*, no. 489 (23 August 1958), 2–25.

30. Baleine, "Conquête," *PM*, no. 231, August 22, 1953, 26.

31. "Von Braun, l'inventeur du V2," *PM*, no. 462, February 15, 1958, 36–38.

32. "Le retour de l'enfant prodige," *PM*, no. 492, September 13, 1958, 66–69.

33. Mitchell R. Sharpe, A bibliography of Wernher von Braun with selected biographical supplement 1930–1969, NASA, http://history.msfc.nasa.gov/vonbraun/vbbiblo.html#ARTICLES-OL (accessed January 28, 2012). The official von Braun bibliography lists three articles by him in *PM*. In fact, at least five pieces in addition to brief interview responses appeared.

34. Quoted in Stella Hughes, "The Apollo Gods with a Devil's Past," *Times Higher Education*, April 18, 1997, http://www.timeshighereducation.co.uk/story.asp?storyCode=101252&sectioncode=26 (accessed December 22, 2011).

35. Wernher von Braun, letter to *PM* editors, April 26, 1966. For further information, see Michael J. Neufeld, *Von Braun: Dreamer of Space, Engineer of War* (New York: Knopf, 2007), 408–9, 535–36 n. 48. Although the letter is quoted in Ernst Stuhlinger and Frederick I. Ordway III, *Wernher Von Braun: Crusader for Space* (Malabar, FL: Krieger, 1996), the original is missing, preventing the checking of the quote's accuracy.

36. "Le roman du conquérant," *PM*, no. 1466, July 1, 1977, cover. The cover photograph showed French rock star Johnny Hallyday; the chronicle of von Braun's life anointed him as the conqueror of the moon.

37. John Gaffney and Diana Holmes, *Stardom in Postwar France* (Providence, RI: Berghahn, 2007), 3–12.

38. Richard Kuisel, *Seducing the French: The Dilemma of Americanization* (Los Angeles: University of California Press, 1993).

39. *PM*, no. 844, June 12, 1965, cover. *Paris Match* stated that it had obtained exclusive rights of coverage for the French-speaking market until the completion of a Moon landing. Although it stated this agreement was penned with NASA, small-print copyright acknowledgments clearly credit Time, Inc.

40. Roger Cartier, "Le premier homme autour de la terre," *PM*, no. 628, April 22,1961, cover; "Titov: près de 700.000 km dans l'espace," PM, no. 645, August 19, 1961, 22–39.

41. "L'homme de l'espace est un homme comme les autres," *PM*, no. 644, August 12, 1961, 21–23.

42. For a clarification of Gagarin's difficult position with the media, see Slava Gerovitch, "The Human Side of a Propaganda Machine: The Public Image and Professional Identity of Soviet Cosmonauts," in *Into the Cosmos: Space Exploration and Soviet Culture*, ed. James T. Andrews and Asif A. Siddiqi (Pittsburgh, PA: University of Pittsburgh Press, 2011), 77–106; Andrew Jenks, "Yuri Gagarin and the Search for a Higher Truth," *Into the Cosmos*, 107–32.

43. Asif A. Siddiqi, "Cosmic Contradictions: Popular Enthusiasm and Secrecy in the Soviet Space program," in *Into the Cosmos*, 47–76.

44. Alexei Leonov, "Mes vingt minutes dans l'espace," interview granted to Marc Heimer, *PM*, no. 835, April 10, 1965, 74–79.

45. *PM*, no. 632, May 20, 1961, cover, 62–71.

46. "Le merveilleux week-end des papas de l'espace," *PM*, no. 845, June 19, 1965, 52–79.

47. "L'espace a fait des vrais jumeaux," *PM*, no. 859, September 25, 1965, 44–49.

48. *PM*, no. 842, May 29, 1965, 56–64.

49. *PM*, no. 842, May 29, 1965, 56–64.

50. Graham Allen, *Roland Barthes* (London: Routledge, 2003), 37.

51. Paul Mathias and Marc Heimer, "Gemini X: Un triomphe, mais pourquoi plus un russe dans l'espace depuis seize mois?" *PM*, no. 903, July 30, 1966, 16–18.

52. See the Gemini coverage in *PM*, no. 912, October 1, 1966, 68–75; no. 920, November, 26, 1966, 42–56; no. 921, December 26, 1966, 92–96.

53. "Ces yeux ont vu la lune de près," *PM*, no. 1032, February 15, 1969.

54. Bernard Giqel, "Les martyrs de l'espace," *PM*, no. 931, February 11, 1967, 24–31.

55. "La mort surprenante du premier héro de l'espace," *PM*, no. 991, April 6, 1968, cover.

56. "Marc Heimer connaissait Komarov," *PM*, no. 943, May 6, 1967, 64–71.

57. "La mort tragique des trois cosmonautes de Soyouz. L'espace les a tués," *PM*, no. 1157, July 10, 1971, cover.

58. "Pas de retard pour la rencontre spatiale russo-américaine," *PM*, no. 1351, April 19, 1975.

59. Bernard Chabbert, *Les fils d'Ariane* (Paris: Plon, 1986), 140–41.

60. Chabbert, *Les fils d'Ariane*, 140–41.

61. Commission générale de terminologie et de néologie, "Répertoire terminologique: astronaute; arrêté du 20 février 1995," *Journal officiel*, September 22, 2000, http://franceterme.culture.fr/FranceTerme/recherche.html, entry "astronaute"(accessed January 22, 2012).

62. The first coverage of Baudry's selection and training appeared in *PM*, no. 1867, March 8, 1985. See also Benoît Clair and Patrick Baudry, *Aujourd'hui le soleil se lève 16 fois* (Paris: Carrère, 1985).

63. Contextually, it is important to consider that France only had two TV channels until 1974 and that the French government did not give up its control monopoly until 1981.

64. Jean Loup Chrétien, *Sonate au clair de terre: Itinéraire d'un Français dans l'espace* (Paris: Denoël, 1993), 10, 69.

CHAPTER 6

> They May Remake Our Image of Mankind:
> Representations of Cosmonauts and Astronauts in Soviet
> and American Propaganda Magazines, 1961–1981
>
> Trevor S. Rockwell

The first human spaceflights inspired visions of humanity transformed by exploring, colonizing, and thriving in space. Some even wondered, as American poet Archibald MacLeish did upon witnessing the first human orbit of the Moon in December 1968, if human space exploration "may remake our image of mankind." To MacLeish, a new notion of humanity's place in the universe had taken shape "in the minds of heroic voyagers who were also men."[1] A half century since humanity pushed one of its own out of this Earthly nest, there seems to be a pervasive sense of disappointment that such visions have gone unfulfilled. As Walter A. McDougall wrote in 2007, "any global consciousness or Spaceship Earth mentality inspired by astronautics has worked no metamorphosis in national or international affairs."[2] But even as these visions of the human future transformed fade into the past, our increasing knowledge of how space exploration narratives were deployed as cold war instruments compels us to take a second look.

This chapter examines how astronauts and cosmonauts were represented in two monthly propaganda magazines between 1961 and 1981. The U.S. Information Agency (USIA) produced the Russian language *Amerika Illiustrirovannoye* (America Illustrated) for distribution in the Soviet Union, and the Soviet Embassy in Washington, DC, published the English language *Soviet Life* for distribution in the United States.[3] The first issues appeared on Soviet and American newsstands in October 1956. As flagship publications, both were attractive, visually stimulating, large-format, glossy magazines printed on high-quality paper and featuring many black-and-white and full-color images. By all accounts, *Amerika* was far more popular with its intended audience as long lines often formed to purchase the magazine. *Soviet Life* never achieved similar appeal with the American public.[4] As official publications, they provide excellent insight into how the Soviet and American governments communicated their visions of space exploration with each other's publics.

The superpowers clearly saw the value of human spaceflight for publicizing their core values to all humankind. Official propaganda on both sides embraced the prediction that space exploration would profoundly refashion humanity. They cast their space explorers as symbols of what they hoped that transformation would bring. Both magazines strongly associated their space explorers with peace and progress. The shared emphasis on peace was likely a practical response to the perils of nuclear war. Progress was one of the central ideals of the Enlightenment, the philosophical movement to which both the United States and the Soviet Union traced their political heritage. Differing conceptions of peace and progress, however, caused the two countries' representations of space explorers to diverge. Whereas *Soviet Life* cast cosmonauts as symbols of peace and progress because of their close ties with socialism and the Soviet state, *Amerika* claimed astronauts signified peace and progress because they personified American freedom and openness.

MacLeish's remark that human space exploration "may remake our image of mankind" elegantly summarizes how the two magazines cast space explorers as heroic humans and personifications of peace and progress who would ultimately transform the human experience. This chapter first discusses how they—the cosmonauts and astronauts—were portrayed in the two magazines. Second, it examines how both publications expected the space explorers to remake humanity. Finally, it looks at how official representations of astronauts and cosmonauts were colored by the divergent images of mankind held by the Soviet and American governments.

"They . . ."

Both *Soviet Life* and *Amerika* routinely portrayed space explorers as embodying both extraordinary heroic and ordinary human characteristics. Employing narratives of danger, courage, and sacrifice, both magazines used representations of heroic space explorers to portray the heroic nation. Hero narratives, in particular, underscored the grand significance of human spaceflight and suggested that these feats had captured the attention of the world. In their extensive coverage of the many celebrations and ceremonies exalting space explorers as heroes, both magazines asserted that these heroes personified national ideals and routinely associated them with their nation's political leaders and institutions. Such a nexus between heroic individuals and their nations implied that leadership in space translated to present and future national leadership on Earth.

Yuri Gagarin's inaugural human spaceflight on April 12, 1961, set the stage for Soviet veneration of cosmonaut heroes. His milestone voyage shifted the emphasis of Soviet space propaganda from robotic to human exploration.[5] *Soviet Life* not only provided detailed accounts of the mission but also covered the many postflight celebrations honoring the first cosmonaut. Its coverage often echoed Soviet domestic propaganda, which declared that the cosmonauts were symbols of the "triumph of socialism."[6] A celebration of Gagarin held on April 14 in Red Square became the blueprint for honoring subsequent cosmonauts. An embrace between Gagarin and Soviet leader Nikita Khrushchev on the roof of Lenin's Mausoleum also initiated a tradition of striving publicly to associate the Soviet leadership with the heroic cosmonauts.[7] With later flights, *Soviet Life* similarly gave extensive coverage to the events held to honor cosmonauts.[8] These articles strongly associated the space heroes with Soviet officialdom via quotations from distinguished speakers and photographs lush with Soviet emblems. Recurrent images of cosmonauts dressed for the formalities in their military uniforms cemented the link between the heroes and the state.[9] Even articles on the Communist Party's 22nd Congress in the fall of 1961 highlighted the participation of cosmonauts Gagarin and Gherman Titov.[10] Other articles regularly associated the cosmonauts with other Soviet social institutions, such as schools or the Young Pioneers.[11]

Amerika pursued a similar tack, routinely reporting on the media coverage of human spaceflight launches, as well as astronauts' postflight celebrations and meetings with the press. The magazine showcased the freedom of the American press by highlighting the American space program's openness to the media. It frequently emphasized, for example, how American space preparations and flights were conducted "in full view" of the world's press and often reported on how American space launches were "swarming with reporters and photographers" from across the world.[12] Images of the American astronauts greeting members of the press were characterized by their informality and by the presence of cameras.[13] Alongside frequent images of astronauts being feted in ticker tape parades, these portrayals linked the American free press to the heroism of the space explorers.[14]

Routine coverage of postflight events, where the American astronauts were bestowed official accolades, established strong links between the heroic space explorers and the nation. A wide-angled shot of the Apollo 11 astronauts' September 16, 1969, address before a joint session of Congress reinforced the link between the astronauts, political institutions, and cherished national symbols by centering the eagle adorning the ceiling of the House chamber. The accompanying text quoted House Speaker John McCormack's statement that the astronauts "represent the best in America."[15] One common pose showed astronauts (and sometimes cosmonauts) in front of the Capitol Building in Washington, DC.[16] Even more common were photographs of astronauts ritually making phone calls to, or shaking hands with, the president. Often still wearing their flight suits, the astronauts thus made contact with the president at the culmination of their heroic journeys.[17]

Both magazines frequently employed danger narratives describing the hazards of human spaceflight—from the uncomfortable to the life threatening. These highlighted space explorers' heroism by focusing on the great physical and psychological demands that spaceflight posed and by typically discussing how only the "best men" were suitable candidates for spaceflight. Danger narratives often also highlighted the many other heroes at work behind the scenes to ensure the space explorers' safety. They thus helped transfer the space explorers' heroism to the broader society.

Both publications also emphasized the space explorers' courage as they responded to the great demands and grave dangers posed by spaceflight. They repeatedly used keywords like calm, comfortable, relaxed, businesslike, and professional to depict space explorers as skilled and capable, performing flawlessly under great duress, and handling adversity with smooth poise. Such an emphasis enhanced their heroic representations of cosmonauts and astronauts while implying that these character traits were as much national as they were individual.

For *Soviet Life*, which was far less open to discuss a mission's difficulties, the training centrifuge commonly served as a symbol of the difficulties and dangers the cosmonauts faced.[18] *Amerika*, in contrast, routinely emphasized the perilous journey into space. In doing so, it furthered the suggestion that the American media enjoyed open access to National Aeronautics and Space Administration (NASA). The magazine openly acknowledged Gus Grissom's "unforeseen accident" aboard *Liberty Bell 7*, the "unexpected crisis" John Glenn faced aboard *Friendship 7*, Scott Carpenter's "continuously troubled flight" aboard *Aurora 7*, and the failure of the automatic reentry systems during Gordon Cooper's *Faith 7* voyage.[19] Its articles on human space missions almost always featured photographs of the astronauts' recovery by helicopter after splashdown.[20] *Amerika*'s frequent discussion of the Mobile Quarantine Facility used after Apollo 11 underlined the unknown dangers of human space exploration. It also demonstrated the American government's concern for the safety of all humankind by showing its effort to, as one article put it, "protect the world from contamination by any possible 'moon germs.'"[21]

The treatment of space explorers' fatalities was a key part of both magazines' danger narratives. Perhaps befitting his prominence as the first cosmonaut, *Soviet Life* gave far more attention to Yuri Gagarin's March 27, 1968, death in a training jet crash than either magazine gave to the seven other space explorers who perished in the line of work between 1967 and 1971.[22] When cosmonauts or astronauts died, memorials in both magazines portrayed their deaths as courageous "sacrifices" made in pursuit of "human progress." They accentuated the fallen space explorer's heroism by predicting the immortality of their names and by noting that their sacrifice had made them "a special kind of human."[23] By pointing out space explorers' ultimate vulnerability to the perils of space travel, danger narratives thus emphasized both sides of the heroic human duality.

Both publications strove to emphasize that their space explorers were not just heroes but also ordinary human beings. Showing the human side of the space explorers strengthened their bonds with their collective societies while increasing their global appeal by connecting them to all of humanity. These narratives of ordinariness built upon a recurrent motif of unity positing that "all mankind" shared in the space explorers' heroic journeys. They thus implied that the world united behind the leadership of the space-faring nations.

In both publications, photographs of astronauts and cosmonauts universally showed them smiling, a detail that made them appear more likeable, friendly, and appealing and also added to their peaceful image.[24] In the same vein, both magazines also highlighted the space explorers' senses of humor, often portraying them joking among themselves or with others.[25] *Amerika* also portrayed the astronauts having fun in space, describing them bounding like kangaroos on the Moon's surface, or taking a playful approach to eating in zero gravity. To underline the human connection, *Amerika* even explicitly made the point that the Apollo 11 astronauts on the Moon's surface were actual men and not "toys"—as they may have appeared on television.[26] The reference to live broadcasts of the moonwalk also reminded readers that they did not share in the historic event. On Soviet television, only the splashdown was broadcast live.

To accentuate the space explorers' ordinariness, both publications commonly focused on their emotional responses to spaceflight, including their awe at the sensory experiences and philosophical implications of their journeys. Routine portrayals of astronauts and cosmonauts being too excited to sleep or eat further accentuated their human nature by emphasizing the necessities of human life. Regular discussions of their loneliness in space likewise made them appear more normal. Loneliness narratives further underlined their connection with humanity back on Earth since the mentioned remedies for loneliness almost always included communications with colleagues or thoughts of family and friends at home.[27]

Another important aspect of both magazines' ordinariness narratives focused on portraying the space explorers' families. Images in *Soviet Life* far more frequently showed the cosmonauts at home with their families while accompanying texts emphasized their dedication to home life.[28] Such a focus demonstrated how representations of Soviet cosmonauts purposefully illustrated the Moral Code of the New Soviet Man, a unique aspect of Soviet social planning to be discussed later. *Amerika* concentrated more on showing how the astronauts' families shared in the spaceflight experience, most often by showing them watching their husbands' and fathers' space journeys on television.²⁹ That such sharing signaled the openness of American society was underlined in a photo in the August 1962 issue showing astronaut Scott Carpenter lifting his son Robin to look into the open hatch of the *Aurora 7* capsule, which was, according to the caption, "on display near the launch site."³⁰

Portrayals of space explorers' families and home life accentuated their ordinariness. But highlighting their traditional roles as fathers and husbands also underlined the space explorers' masculinity, an emphasis both magazines routinely made. *Soviet Life*, for example, frequently depicted the cosmonauts in training or in other lifestyle settings that showed off their muscular physiques.³¹ *Amerika* did the same but showcased the astronauts' masculinity in other ways too. It often charted the growth of the astronaut's beards during their space voyages, for example.³²

Of course, not all cosmonauts in this period were men. The June 1963 flight of the world's first female space traveler, Valentina Tereshkova, set another milestone for Soviet "manned" spaceflight. She was the only female cosmonaut to fly until nineteen years later. Despite this fact, Soviet propaganda proclaimed her flight, as Chief Designer Sergei Korolev described it to Soviet journalists, "one of the most striking demonstrations of the equality of Soviet women."³³ As much as her male colleagues performed as models of the New Soviet Man, Tereshkova symbolized a New Soviet Woman. Her representation was thus unique among Soviet cosmonauts. In *Soviet Life*, her image essentially retreated over time from the heroic domain to the human one, as her role as a wife and mother eventually displaced her role as a cosmonaut. According to an August 1963 article reporting on her spaceflight: "Valya Tereshkova is really two girls: one in orange coveralls, the other in a sky blue dress."³⁴

Soviet Life's representations of her soon exchanged her orange coveralls for civilian clothes. A photograph taken of the Soviet cosmonauts atop the Lenin Mausoleum appearing in the January 1964 issue pictured her both among her male colleagues and set apart from them by her attire. She was the only one in the group not wearing a military uniform (Figure 1).³⁵ Elsewhere in the same issue she wore a bridal gown, as the magazine celebrated her marriage to fellow cosmonaut Andrian Nikolayev.³⁶ A playfully titled article in October 1965 illustrated how giving birth to a "cosmotot" signaled the transformation of her identity from cosmonaut to "cosmonette."³⁷ By October 1970 the previously exalted first woman in space was no longer portrayed as a cosmonaut and not even referred to by her well-known name. As the caption to a photo of her with

FIGURE 1. Cosmonauts (from left to right) Pavel Popovich, Andrian Nikolayev, Gherman Titov, Valentina Tereshkova, Yuri Gagarin, and Valery Bykovsky atop Lenin's Mausoleum in Red Square for a June 22, 1963, celebration honoring the recent spaceflights of Tereshkova and Bykovsky. Note their friendly smiles, military attire, and how Valentina Tereshkova is set apart from the men by her civilian clothes. The photograph appeared in Oleg Ivanov, "Cosmonaut's Joke—Photo of the Month," *Soviet Life*, January 1964, 37. Courtesy ITAR-TASS News Agency/Vladimir Savostyanov; Vasily Yegorov.

daughter Elena indicated, she was now simply "Andrian Nikolayev's wife."[38] Emphasizing Tereshkova's domestic roles reflected a Soviet retreat from gender egalitarianism as traditional gender divisions resurged during the Brezhnev period.[39] Even at the height of her fame, when she was predominantly cast as a symbol of Soviet egalitarianism, there was a wide gulf between official propaganda about her and the sexism she and the other female cosmonauts faced behind the scenes.[40]

". . . May Remake . . ."

Both the United States and the Soviet Union were born in revolutions. At their core, the foundational ideals of both were based on a belief in profound social transformation. In turn, this revolutionary impulse emerged from the Enlightenment of the eighteenth century, when European philosophers sought to apply scientific reason to reform society in the name of progress. In the twentieth century, as the two superpowers took to space, their propaganda expressed a shared vision of the power of human space exploration to transform humanity. They both frequently described human spaceflight as a turning point in history and the beginning of a new era. They related it to epochal shifts in human evolution and presented it as the next step in that evolution. In doing so, they both predicted an ambitious future of a broad human presence in space that would revolutionize the human experience. The two nations had differing ideas about *how* humankind would be transformed, and these divergent values shaped the two magazines' representations of humans in space.

The frequency and descriptive detail of these visions of the future of human spaceflight reached a climax in both magazines in the late 1960s and early 1970s.[41] In *Amerika*, the peak came immediately following Apollo 11. Despite Nixon's call for a "bold" yet "balanced" approach to space exploration in the 1970s, the magazine eschewed balance to emphasize a highly ambitious vision of humankind's future in space, led by the United States.[42] It thus sought to capitalize on the attention garnered by the historic lunar landing to propose that the United States would continue to make "giant leaps for all mankind" for generations to come.[43] Its representations of future astronauts implied that the free and open flow of ideas was essential to bring about peace and progress. Routine articles on the prospects of human spaceflight promised great things, and their realism and detail contrasted American openness against the secrecy of the Soviet space program.[44]

Soviet Life's images of the future were far less concrete, far more idealistic, dreamlike, and fantastical than American ones, but no less transformative and no less convinced that the future would bring a vast expansion and intensification of the human presence in space. Instead of giant leaps, the magazine emphasized a graduated and "multipurpose" Soviet approach based on "consistency" and "progress by stages."[45] As late as July 1968, it still suggested that a piloted Soviet lunar landing was in the works.[46] But beginning with the August 1969 issue, which applauded Apollo 11, a series of articles put forward the idea that Soviet planners had long since decided that robotic probes were superior for exploring the Moon and other bodies.[47] In spite of this argument and the increased verve with which the magazine celebrated Soviet robotic probes during these years, *Soviet Life* never abandoned putting humans front and center in its narratives of present and future exploration of space.[48] Indeed, with its Soyuz series of missions, the Soviet space program had maintained a relatively continuous human presence in Earth orbit after late 1968. Even in its coverage of robotic probes, *Soviet Life* routinely noted how the ability to perform a "soft landing" was a significant step toward human voyages to the Moon, Mars, Venus, and beyond.[49] It also focused on portraying space stations functioning as "spaceports" for interplanetary human voyages.[50]

Narratives of an imminent and grand human presence in space were critical to justifying both sides' routine propaganda claims that human spaceflight milestones marked turning points ushering in a new era in history.[51] Predictions of a pronounced transformation of humanity increased the significance of past and present achievements and the efficacy of the propaganda that celebrated them. As the living, breathing actors turning the page of this history, cosmonauts and astronauts were integral to both magazines' narratives of how space exploration would remake humanity. As both publications rhetorically emphasized, these explorers were simply the *first* humans to venture into space. As "pioneers" exploring a "new frontier" they broke a path that all humanity was inevitably, both publications enthusiastically predicted, soon to follow.[52]

Both magazines used representations of space explorers to illustrate how their values would lead humanity through the profound transformative process underway. Adopting many of the rationales and assumptions of space advocates, both publications' narratives of the future argued emphatically that human spaceflight would bring many positive benefits and would greatly contribute to peace and progress on Earth.[53] These too echoed Enlightenment beliefs that linked scientific discovery with social advancement.

Strongly associating space explorers with science served both the peace and progress themes. Both publications claimed science to be an inherently peaceful activity and emphasized its international character. These assumptions underscored their nation's leadership of the world scientific community and presented space activities as a benefit to all humanity. Both magazines described human spaceflight deriving incalculable benefits, including new knowledge, new resources, new industries, and new technologies. They both also argued that a nation's scientific, technological, and economic capabilities were essential for exploiting the potential boon of human spaceflight. Representations of cosmonauts and astronauts played a key role in demonstrating these capabilities.

Amerika commonly used science to justify its claims that human space exploration was vital by arguing that human operation and oversight of space experiments made for better science. It frequently discussed the scientific training and activities of American astronauts. Some of the most prominently placed images drew attention to American astronauts wielding scientific instruments on the Moon. It often described the astronauts' efforts to obtain lunar specimens, which served as important symbols of American openness in repeated discussions of how they were shared with the international scientific community on Earth. The magazine's invariable emphasis on sharing other scientific data and detailed reports of the United States' current and future scientific space missions likewise underscored the notion of American openness.[54]

Although *Soviet Life*'s articles on robotic lunar probes in the late 1960s argued that such mechanistic exploration of space made for safer and cheaper science, over the longer term the magazine strove to depict cosmonauts' key role in scientific aspects of the Soviet space program. It frequently emphasized their scientific training, invariably highlighted the scientific experiments they performed in space, and sometimes even referred to them as "scientist-cosmonauts."[55] *Soviet Life* was far more nebulous than its American counterpart, however, about the precise nature of the science performed during Soviet human spaceflights.

Cosmonauts frequently discussed vaguely making observations, performing experiments, and collecting data, but without further detail.[56]

To fully depict human spaceflight as a peaceful endeavor, it was vital for space explorers and their nations to be seen embracing a cooperative spirit. According to the two magazines, this would prevent earthly tensions from spreading into space and would reduce duplication of effort as humanity faced the great expenses and challenges of exploring the planets and beyond. Both magazines expressed support for cooperation on human spaceflight long before the highly symbolic July 1975 Apollo-Soyuz Test Project (ASTP) joint mission was conceived. Over several years of ongoing dialogue before the May 1972 agreement for a joint flight, both publications explicitly endorsed space cooperation and reported favorably on the more limited joint agreements not involving direct collaboration between cosmonauts and astronauts.

Just as astronauts and cosmonauts provided the vital hands for ASTP's emblematic "handshake in space," they also played a key role in opening the door to cooperation, through their professional contacts and friendly relations over the years. Throughout the 1960s and 1970s, whenever astronauts and cosmonauts visited each other's countries, both magazines invariably covered their exchanges and portrayed them as friendly meetings.[57] These reports encouraged the further deepening of professional ties since these exchanges were typically portrayed as strengthening peace or increasing "mutual understanding." Many articles highlighting their joint preparations for ASTP cast them in central roles as ambassadors, colleagues, and friends.[58]

Over the years of discussion preceding the ASTP agreement, American officials consistently took the lead in pursuing cooperation in human spaceflight while becoming increasingly concerned that Soviet secrecy would preclude doing so.[59] Openness was a key issue in the effort to undertake a joint human spaceflight since participation in such a mission would require the Soviets to offer American astronauts and technicians unprecedented access to their space program. Drawing attention to American openness, *Amerika* sometimes mentioned how visiting cosmonauts had been invited to visit NASA facilities but had politely declined since they could not reciprocate by inviting astronauts to Baikonur.[60] Even during ASTP preparations, as Roald Sagdeev has reported, Soviet officials undertook deceptions to prevent American officials from perceiving the Soviet space program's relationship with the military establishment. Soviet military officers overseeing the space program wore civilian clothes, for example, and falsely claimed to defer authority to the Academy of Sciences.[61]

ASTP was one of the most prominent symbols of détente.[62] Reflecting this political context, both publications downplayed the ideological differences and animosities between the two countries in their coverage of the joint flight and the nearly three years of preflight preparations. They concentrated instead on certain key "compatibility issues," such as language, air pressure differences between the two space capsules, and the docking mechanism. The language issue especially highlighted the human dimension of the mission and put the crew front and center in the overarching narrative of finding "mutual understanding." The reciprocal agreement that cosmonauts would use English and astronauts use Russian during the flight was repeatedly noted. The language issue, like the handshake, cast individual space explorers as symbols of their countries' willingness to overlook ideological differences and build a reciprocal relationship. In such a symbolic framework, improvements in interpersonal relations between astronauts and cosmonauts were portrayed as significant steps toward the advancement of human progress and the achievement of global peace. To this end, both magazines showcased the ASTP crew's comments on how their individual relations became closer over time and especially highlighted those remarks that hinted at such an analogy between the interpersonal and international dimensions.[63] Representations of cosmonauts and astronauts were thus fundamental to demonstrating the scientific and cooperative, peaceful, and progressive nature of each nation's space program.

". . . Our Image of Mankind"

The core beliefs and values of the two superpowers formed and shaped their image of humanity and of human spaceflight. Both magazines' emphatic use of the peace and progress motifs points to some commonalities in the Soviet and American historical trajectories. The two countries' political ideals shared a common ancestor in Enlightenment notions of progress achieved through applying scientific reason to reform society. In the cold war, they were both also the key agents of humankind's potential "mutually assured destruction" in a nuclear conflict, hence their desire to underline a message of peace. Both sides strongly associated human spaceflight with peace and progress, but their deviating definitions of those ideals led them to differing representations of cosmonauts and astronauts. Divergent ideologies thus shaped the way that the two magazines depicted space explorers remaking "our image of mankind."

Soviet Life's conceptions of peace and progress were distinctly defined in relation to socialism. In Soviet rhetoric, peace and socialism were considered indivisible. In Soviet discourse, the "struggle for peace" thus equated with the "struggle for socialism," and the goal of "world peace" similarly meant "world socialism."[64] "Peaceful coexistence" was perhaps the most prominent slogan of Soviet propaganda and international discourse during the first few decades of space exploration.[65] Soviet officials repeatedly clarified that the term did not lessen the ideological struggle against capitalism. On the contrary, it meant that the struggle should shift from the military arena to economic, cultural, scientific, and technological forms of competition.[66] Capitalism—frequently labeled "imperialism"—was considered the principal enemy of peace, and the Soviet state claimed to protect Soviet citizens from the "imperialist aggressor," which often meant "American aggression" and "militarists."[67]

Both space exploration and propaganda played prominent roles in the struggle for peaceful coexistence. In Soviet propaganda about space exploration, cosmonauts thus played a vital role in projecting the regime's commitment to peace. *Soviet Life* explicitly linked Yuri Gagarin's historic spaceflight with "the sacred cause of peace." Quoting Khrushchev, the magazine explicitly declared that it illustrated the "conditions created by the October Socialist Revolution." *Soviet Life* strongly associated the flight with socialism and peace and suggested that other nations were intent on militarizing space.[68] It commonly showed cosmonauts in their official military uniforms. In the symbolic parlance of the Soviet Union, however, such military attire did not identify the cosmonauts with war, but rather with the Soviet state and its defense of peace.[69] An often-reproduced photo appearing in the September 1962 issue of a smiling, uniformed Gagarin holding a dove made the point less obliquely, as did illustrations elsewhere featuring rockets and doves.[70]

At the same time, socialism, as a postcapitalist order, also implied progress. Central to Soviet myth were the related ideas that the socialist state had made a dramatic break with the capitalist and imperialist human past and that it promised even more profound changes as it reached forward to the goal of building communism.[71]

Representations of cosmonauts were modeled on a "Moral Code of the Builder of Communism" that the Communist Party of the Soviet Union unveiled just six months after Gagarin's flight.[72] Reading like a blueprint for how cosmonauts were to be represented in propaganda, the Moral Code called for the New Soviet Man to be committed to Soviet progress toward building communism, conscientious for the good of society, honest, morally pure, modest, respectful family members, and especially concerned for raising their children with these same values.[73] Yuri Gagarin's autobiography, *The Road to the Stars*, was a key text that set the tone for how cosmonauts were to personify the New Soviet Man.[74] Between July and September 1961, *Soviet Life* published several lengthy excerpts from *The Road to the Stars* as part of its coverage of Gagarin's historic spaceflight.[75] Overall, the magazine's portrayals of Gagarin, and subsequent cosmonauts, conformed to the tenets of the Moral Code, which called for them to appear as model socialists and to represent the Soviet commitment to peace and progress.

Just as Soviet human spaceflights were declared triumphs of socialism, so were cosmonauts cast as socialist heroes. Frequently pointing out the cosmonaut's proletarian heritage, *Soviet Life* routinely identified them with the working class.[76] Their accomplishments, as specifically socialist achievements, were presented as a source of pride in socialism. Their behavior, similarly portrayed as socialist behavior, provided a socialist model to be admired, respected, and perhaps emulated. Depicting the cosmonauts as model socialists thus presented them to signify that the Soviet Union was best suited for guaranteeing peace and progress as space exploration refashioned the human future.

The Eisenhower administration had termed the Soviet emphasis on improving relations with the West a "peace offensive."[77] In the late 1950s USIA officials began to search for a response to peaceful coexistence.[78] In June 1961, USIA Director Edward R. Murrow and his associates articulated—and received President Kennedy's blessing for—an "effective countertheme" based on identifying the United States with an "open world" and a "world of free choice."[79] In contrast to *Soviet Life*, which connected cosmonauts with the strong socialist state, *Amerika* presented astronauts as representatives of an American spirit characterized by the ideals of freedom and openness used to differentiate an American conception of peace from the Soviet concept of coexistence. In this view, the United States' human spaceflight achievements signified nothing less than the American spirit's ascendancy into space.

Central to *Amerika*'s portrayals of astronauts was the motif of a free market, not only the economic market but also the marketplace of ideas, as the magazine routinely demonstrated the astronauts' gratification with life in a free society. For example, *Amerika* often showcased American astronauts' freedom to worship. Frequent depictions of American astronauts performing religious rites, such as prayer, communion, and reading from the Bible, in outer space demonstrated openness by highlighting religious tolerance.[80] Such images contrasted American astronauts with the Soviet cosmonauts, who Soviet propagandists cast as symbols of atheist scientific materialism. *Amerika* thus targeted those in the Soviet Union who still turned to religion in spite of the atheistic propaganda aimed at them.[81] Astronauts' expressions of faith also supported space exploration rhetoric about the unity of "all mankind" to underline the associations between exploring the cosmos and preserving peace on Earth. In one piece, for example, Edwin "Buzz" Aldrin Jr. described how, while serving himself communion on the Moon, he prayed that people would "see the deeper meaning" behind space exploration and would "recognize that we are all one mankind, under God."[82] Visuals associating spaceflight with Christianity were also common. Paintings by Robert McCall, which often appeared in the magazine, frequently used the pictorial motif of rays of light in the shape of a cross to invest the celestial "heavens" with spiritual content.[83]

Historian Kenneth Osgood has described how the USIA generally promoted to foreign audiences the American way of life: a celebration of freedom and democracy, especially of the free market economy, which brought prosperity and a myriad of goods within easy reach of U.S. consumers.[84] In line with this, *Amerika* often showcased the material abundance that astronauts and their families enjoyed and implied that such affluence was a product of the free market.[85] Recurrent images of Americans witnessing human spaceflight on television sets in their homes or in front of large screens set up in public signified that affluence and leisure time were abundant in American society.[86] So did recurrent images of astronauts (and cosmonauts) at Disneyland and Disneyworld. One especially conspicuous photo essay covering the October 1969 U.S. tour of Georgi Beregovoi and Konstantin Feoktistov showed the cosmonauts, among other things, barbecuing steaks in San Diego, riding in an experimental car in Detroit, and wearing Mickey Mouse caps at Disneyland.[87]

Articles on daily life for astronauts in space routinely focused on the material comforts American astronauts enjoyed in orbit and especially highlighted those items most suggestive of affluence and leisure. Skylab astronauts, for example, had several hours each day to "relax and enjoy" such things as personal portable tape recorders.[88] Long, detailed descriptions of space menus including such delectable entrées as filet mignon and lobster Newburg were routine.[89] Wide-angle photographs of astronauts' families watching human spaceflights on television or of lavish banquets honoring the astronauts showcased the material prosperity of American consumers: fashionable clothing, hairstyles, furniture, modern homes, television sets, and a bounty of food, drink, and cigarettes.[90] Among these, recurring pictures of astronauts and their family members with open mouths may have been intended to suggest openness (Figure 2).[91] Such images of openness and freedom symbolized the American conception of peace, whereas the material abundance on display strongly indicated industrial progress.

Soviet Life similarly used cosmonauts to showcase industrial progress and to suggest that the socialist system provided an abundance of material goods. However, since the Soviet economy provided relatively far fewer consumer items, it by necessity did not depict them enjoying those comforts in the home. Instead, it showed visiting American astronauts enjoying official Soviet hospitality or touring Soviet industrial sites.[92] Although it often rhetorically linked space accomplishments with economic progress, *Soviet Life* only very occasionally used cosmonauts to showcase the availability of consumer goods in Soviet society.[93] It rarely discussed space (and Earth) food in a manner to suggest abundance.[94] One article exhibited several ASTP souvenir items, including "space

FIGURE 2. The Gemini 4 astronauts and their wives: (from left to right) Patricia White, Ed White, Patricia McDivitt, and James McDivitt at a press conference held at the Manned Spacecraft Center in Houston, Texas, on June 11, 1965, attended by President Johnson. Note their civilian attire, friendly smiles, open mouths, and the presence of their family members. The photograph appeared in "Happy End to a Successful Space Flight," *Amerika*, November 1965, 34–35. Photo by Francis Miller/Time Life Pictures/Getty Images.

cigarettes," chocolates, perfume, and vodka.[95] Generally speaking, however, images of Soviet material abundance were atypical in *Soviet Life*, which was just as likely to criticize American consumerism as it was to showcase the Soviet version.[96] The Soviet magazine's emphasis on abundance also did not appear until the 1970s, suggesting that the theme emerged in response to the rich and numerous displays of material affluence in *Amerika*.

Conclusions

The similarities between the two magazines examined here are far more striking than the differences. Although both publications articulated different definitions of peace and progress and employed different media strategies, they both portrayed space explorers as extraordinary heroes and ordinary human beings to link them to their nations and all humankind. But more remarkably, the two magazines shared the basic view that human spaceflight would prove powerfully transformative and would ultimately bring peace and progress to humanity. The mutual emphasis on these themes stemmed from the common influence that the Enlightenment and the cold war had on the Soviet Union and the United States. Additionally, the prerogatives of propaganda led both publications to illustrate a bold future to heighten the significance of their current and past achievements in space. They showed how space exploration "may remake our image of mankind" to cast their own nation leading the next step of human evolution. As Neil Armstrong famously observed, the astronauts and cosmonauts did not simply take small steps—they made giant leaps.

This chapter began by noting how human spaceflight initially inspired predictions that stepping into space would forever change humanity and how such heated enthusiasm has more recently chilled into disappointment as we remain (so far) in the doldrums of low Earth orbit. Any attempt to assess whether human spaceflight has fulfilled its initial promise should begin by assuming that there were multiple narratives. Soviet and American official narratives were among those that most enthusiastically imagined space exploration refashioning humankind, and they both did so as part of a broader project of claiming their nation's global leadership. We should not, therefore, ask *whether* human spaceflight has remade our image of ourselves, but *how*.

Notes

1. MacLeish's piece was widely disseminated after its debut on the front page of *The New York Times* on Christmas Day, 1968. Excerpts from it made their way into Richard Nixon's January 20, 1969, inaugural address and into the flagship propaganda publication *Amerika* magazine on more than one occasion. See Archibald MacLeish, "A Reflection: Riders on Earth Together, Brothers in Eternal Cold," *New York Times*, December 25, 1968; U.S. Department of State, *Foreign Relations of the United States, 1969–1976*, vol. 1 (Washington, DC: U.S. Government Printing Office, 2003), 53–55 (hereafter *FRUS* with year and volume number); "Apollo 8: Now Man Has Circled the Moon," *Amerika*, May 1969, 43–46; Archibald MacLeish, "A Reflection (Introduction to Special Section on Apollo 11)," *Amerika*, November 1969, 25.

2. Walter A. McDougall, "A Melancholic Space Age Anniversary," in *Remembering the Space Age*, ed. Stephen J. Dick (Washington, DC: NASA History Division, 2008), 390.

3. The Soviet magazine was called *USSR* from 1956 to 1964 and *Soviet Life* from 1965 to 1991. (For the sake of simplicity this essay refers to the Soviet magazine's entire run as *Soviet Life*, and for the sake of brevity, it refers to *Amerika Illiustrirovannoye* as simply *Amerika*.) The production of *Soviet Life* was overseen by the Soviet Department of Agitation and Propaganda and after April 1961 by the Agentstvo Pechati Novosti.

4. A series of exchange agreements beginning in 1956 governed the two magazines' circulation. Throughout the period discussed here, American officials sought to increase the allowable circulation of the exchange agreement, whereas Soviet officials wanted to limit the distribution of *Amerika* to levels matching the dismal sales of *Soviet Life* in the United States. See Walter L. Hixson, *Parting the Curtain: Propaganda, Culture, and the Cold War, 1945–1961* (New York: St. Martin's Press, 1997), 117; Yale Richmond, *Cultural Exchange and the Cold War: Raising the Iron Curtain* (University Park: Pennsylvania State University Press, 2003), 2, 149–51; Yale Richmond, *Practicing Public Diplomacy: A Cold War Odyssey* (New York: Berghahn Books, 2008), 100–103; Max Frankel, "U.S.-Soviet Accord in Cultural Field Extended 2 Years," *New York Times*, November 22, 1959; "Texts of U.S.-Soviet Accords on Exchanges in Technical and Cultural Matters," *New York Times*, June 20, 1973.

5. Michael J. Sheehan, *The International Politics of Space* (New York: Routledge, 2007), 31.

6. *Pravda* described Gagarin's spaceflight this way. See *Pravda*, April 15, 1961. *Soviet Life* described Vostok as a "new triumph of Lenin's ideas, a confirmation of the correctness of the Marxist-Leninist teachings." See "A Day to Remember," *USSR*, May 1961, 2–3.

7. Khrushchev and Gagarin appeared together on the May 1961 front cover, embracing on the inside front cover, and several more times inside the issue. See "Nikita Khrushchev with Yuri Gagarin and His Wife Valentina Driving along a Street in Moscow on Their Way to Red Square on April 14," *USSR*, May 1961, front cover; "The First Man in Space: Yuri Gagarin," *USSR*, May 1961, inside front cover, 1; "A Day to Remember," 2–3; "Heartfelt Gratitude," *USSR*, June 1961, inside front cover, 1; "Moscow Welcomes the Hero," *USSR*, June 1961, 6–9.

8. For examples, see "Second Soviet Cosmonaut in Outer Space," *USSR*, September 1961, 14–15; Gherman Titov, "435,000 Miles through Space," *USSR*, October 1961, inside front cover, 1–7; "Hero's Welcome for Cosmonauts in Moscow," *USSR*, October 1962, 26–27; "Moscow Welcomes Cosmonauts," *USSR*, August 1963, 26–27; "Interview by Leonid I. Brezhnev, General Secretary of the CPSU Central Committee for French Television," *Soviet Life*, December 1976, 8c; Boris Petrov, "The Space Experiment," *Soviet Life*, January 1969, 10–11.

9. See especially the images of the first six cosmonauts in official portraits wearing their uniforms and many medals in "Do You Know Soviet Cosmonauts?" *Soviet Life*, April 1964, 32–33.

10. "Who Are the Delegates?" *USSR*, November 1961, 3–7; "Congress Delegates Speak," *USSR*, December 1961, 14–16; "Congress Delegates Speak," *USSR*, September 1962, 24–26.

11. See especially Yuri Yakovlev, "Young Pioneers," *Soviet Life*, January 1963, 26–29; "Young Cosmonauts' Club," *Soviet Life*, August 1969, 16–21; "Birthplace of First Spaceman Rebuilt by Students," *Soviet Life*, September 1974, 18–23.

12. Charles Gregory, "As a Nation Watched . . . 'Lift Off!'" *Amerika*, August 1961, 36–39. *Amerika* even explained that the American space program "is based on the profound belief that all are entitled to a full awareness of both our successes and to the difficulties." Jeff Stansbury, "John Glenn . . . in Orbit," *Amerika*, May 1962, 2–7. See also Anthony J. Bowman, "Space Travelers Meet," *Amerika*, August 1962, 2; "Saturn V Takes a Giant Step toward the Moon," *Amerika*, April 1968, 48–49; "00:4; 00:3; 00:2; 00:1; 00:0 (Apollo 11 Lift-Off)," *Amerika*, April 1970, 9.

13. For a particularly telling image of Shepard and Grissom having a laugh and sharing the frame with a photographer who has two large cameras around his neck, see Gregory, "As a Nation Watched," 36–39. See also "Apollo 8: Now Man Has Circled the Moon," 43–46; Jeff Stansbury, "On Target: Flight of Second U.S. Astronaut," *Amerika*, December 1961, 9. Photographs of space explorers making public appearances were commonly framed to showcase the audiences participating at these events, further reinforcing the message that the American media was free and open to the public. See, for example, Bowman, "Space Travelers Meet," 2–3.

14. See the back cover of the May 1962 issue and especially "Welcome Back! (Apollo 11)," *Amerika*, April 1970, 60–61, 63.

15. "Welcome Back! (Apollo 11)," 62–65. For a similar image of John Glenn's 1962 address before Congress, see Stansbury, "John Glenn," 7.

16. Bowman, "Space Travelers Meet," 2; Everly Driscoll, "Apollo Astronauts: Where Are They Now?" *Amerika*, July 1975, 30–31.

17. Stansbury, "On Target," 9; Stansbury, "John Glenn," 7; Bowman, "Space Travelers Meet," 3; John Noble Wilford and James T. Wooten, "To the Moon and Back (Excerpt from Apollo 11: On the Moon)," *Amerika*, April 1970, 6; "Welcome Back! (Apollo 11)," 61–65; "Apollo Soyuz Project: First International Manned Spaceflight," *Amerika*, July 1975, 24.

18. On the training centrifuge, see "Man's First Flight into Space," *USSR*, June 1961, 2–5; Norair Sisokyan, "The Road to the Stars," *USSR*, June 1961, 10; Ilya Kopalin, "First Flight to the Stars," *USSR*, November 1961, 16–19; "Space Merry-Go-Round," *Soviet Life*, August 1966, 54–55; "Space, Interviews with Konstantin Feoktistov and Oleg Gazenko," *Soviet Life*, October 1976, 2–9. Soviet space launches, unlike American ones, were kept secret from both domestic and international audiences. Only successful launches were announced publicly and only after the fact. Setbacks and failures were never publicized. See G. Perry, "Perestroika and Glasnost in the Soviet Space Programme," *Space Policy* (November 1989): 283; Sheehan, *The International Politics of Space*, 32.

19. Stansbury, "On Target," 9; Stansbury, "John Glenn," 2–7; Marjorie Parsons, "'Faith-7' in Space," *Amerika*, September 1963, 11–13; "Pioneers Together—Astronauts and Cosmonauts," *Amerika*, November 1969, 30–31; "A Calendar of Space Flight," *Amerika*, April 1970, 44. Interestingly, the "Calendar of Spaceflight," by focusing only on human missions, neatly avoided recalling the spectacular American launch failures of the late 1950s.

20. A select few examples include "Meeting in Space," *Amerika*, October 1963, 38–39; Ralph Segman, "Gemini: Beginning and Successful Ending of a Project," *Amerika*, October 1967, 34–35; "On the Moon (Apollo 11)," *Amerika*, April 1970, 38–39; Jay Holmes, "The New Configuration," *Amerika*, February 1971, 18; Arthur Pariente, "Apollo 15: Touchdown for Science," *Amerika*, December 1971, 30.

21. Wilford and Wooten, "To the Moon and Back," 2–8.

22. Apollo astronauts Virgil Grissom, Ed White, and Roger Chaffee died on January 27, 1967. Cosmonaut Vladimir Komarov perished aboard Soyuz 1 on April 24, 1967. Soyuz 11 cosmonauts Georgi Dobrovolski, Viktor Patsayev, and Vladislav Volkov died on June 30, 1971.

23. See especially "Their Deeds Will Live Forever," *Soviet Life*, September 1971, 13–16; "Earth Will Remember Them Forever," *Soviet Life*, August 1969, 27; Nikolai P. Kamanin, "His Life's Cause," *Soviet Life*, July 1968, 34; "Yuri Gagarin Citizen No. 1 of the Universe," *Soviet Life*, July 1968, 25–27; Yaroslav Golovanov, "He Wanted to Speed Up History," *Soviet Life*, July 1968, 28–29; Robert Rozhdestvensky, "Continue Their Work," *Soviet Life*, September 1971, 17. For American examples, see Sherwood Harris, "Cape Kennedy: The Moon Has Changed the View," *Amerika*, October 1967, 20–23; "A Calendar of Space Flight," 44–51.

24. See especially the December 1962 front and back covers: "Titov, Gagarin, Nikolayev and Popovich Are Interviewed by *USSR* Staff Writers," *USSR*, December 1962.

25. See especially "Welcome Gherman Titov," *USSR*, June 1962, 6–9; Pavel Bakashev, "Orbit Round the Earth," *USSR*, August 1962, inside front cover, 1; Oleg Ivanov, "Cosmonaut's Joke—Photo of the Month," *Soviet Life*, January 1964, 37. For American examples, especially see Gregory, "As a Nation Watched," 39; Stansbury, "John Glenn," 2, 5; Bowman, "Space Travelers Meet," 3; Parsons, "'Faith-7,'" 12–13; Frank Borman, "A Special Message to the Readers of *Soviet Life*," *Amerika*, June 1969, 40; "On the Moon (Apollo 11)," 30–31, 39; "Apollo Soyuz Project," 22–23.

26. Wilford and Wooten, "To the Moon and Back," 2–8; "Special Report: Man on the Moon," *Amerika*, November 1969, insert between pp. 28 and 29.

27. For Soviet examples, see "Yuri Gagarin's Own Story," *USSR*, May 1961, 4–5; Olga Apanachenko and Vasili Peskov, "The Family Waits," *USSR*, May 1961, 62–63; "Man's First Flight into Space," 2–5; Sisokyan, "The Road to the Stars," 10; Yuri Gagarin, "Road to Outer Space," *USSR*, July 1961, 24–29; August 1961, 7–11; September 1961, 16–23; Titov, "435,000 Miles," 1–7; Alexander Mokletsov, "Lady Cosmonaut on Vacation," *USSR*, December 1963, 22–26; Vladimir Zhukar and Irina Pushkina, "Space Psychology," *Soviet Life*, April 1965, 34; "Space Explorers' Poll," *Soviet Life*, August 1969, 4–8. For American examples, see Stansbury, "On Target," 9; Stansbury, "John Glenn," 2–7; Bowman, "Space Travelers Meet," 2–3; Fady Bryn, "Astronaut Carpenter Lifts Off," *Amerika*, August 1962, 7–9; Parsons, "'Faith-7,'" 11–13; Jim Schefter, "Our First Day on the Moon—What Will It Be Like?" *Amerika*, May 1969, 47–51; "A Day in Outer Space," *Amerika*, November 1969, 27–28; "Here Come the Cosmonauts!" *Amerika*, March 1970, 48; Wilford and Wooten, "To the Moon and Back," 2–7; "Our Impossible Goal," *Amerika*, April 1970, 40; "Special Report," insert between pp. 28 and 29.

28. Apanachenko and Peskov, "The Family Waits," 62–63; "Yuri Gagarin—World's First Cosmonaut," *USSR*, June 1961, 11–13; Gagarin, "Road to Outer Space," July 1961, 24–29; August 1961, 7–11; September 1961, 16–23; Titov, "435,000 Miles," 1–7; "Cosmonaut Andrian Nikolayev," *USSR*, October 1962, 40–41; "Cosmonaut Pavel Popovich," *USSR*, October 1962, 42–43; Anatoli Blagonravov, "Space Spidermen," *Soviet Life*, April 1969, 48; Mikhail Rebrov, "Stellar Town: Their Earthly Home," *Soviet Life*, August 1969, 36–38; "Cosmonauts Town," *Soviet Life*, July 1975, 20–23.

29. Segman, "Gemini," 34–35; "Special Report," insert between pp. 28 and 29; "On the Moon (Apollo 11)," 36–37; "Our Impossible Goal," 40–41; Wilford and Wooten, "To the Moon and Back," 2–7.

30. Bryn, "Astronaut Carpenter," 9.

31. See, for example, "Yuri Gagarin's Own Story," 4–5; Gagarin, "Road to Outer Space," July 1961, 24–29; August 1961, 7–11; September 1961, 16–23; "Spaceman on Vacation," *USSR*, September 1961, 17; Titov, "435,000 Miles," inside front cover, 1–7.

32. See, for example, Bryn, "Astronaut Carpenter," 7; "A Day in Outer Space," 27–28; Wilford and Wooten, "To the Moon and Back," 2–8.

33. G. A. Skuridin, V. I. Sevast'yanov, and G. A. Nazarov, *Entrance of Mankind into Space (15th Anniversary of the First Manned Flight into Space)*, NASA Technical Translation NASA-TT-F-17114 (Washington, DC: NASA Technical Translation Service, 1976), 31. Cited in Sheehan, *The International Politics of Space*, 31.

34. "Seagull Calling, Report from Cosmonaut Vasili Peskov," *USSR*, August 1963, 22–23.

35. Ivanov, "Cosmonaut's Joke," 37.

36. Valeri Gendi Rot, Yuri Korolev, and Alexander Mokletsov, Rot et al., "Marriage Made in Space," *Soviet Life*, January 1964, 38–43.

37. "Cosmonaut, Cosmonette, Cosmotot," *Soviet Life*, October 1965, 48–49. For other articles on Tereshkova, see Mokletsov, "Lady Cosmonaut," 22–26; Valentin Mikhailov, "Lady Cosmonaut's Mailbag," *Soviet Life*, July 1964, 4–5.

38. Vasili Pavlov, "424 Hours in Space," *Soviet Life*, October 1970, 12. This is not to say that *Soviet Life* did not still value the image of an active, heroic, adventurous working female as a symbol of equality between the sexes in the Soviet Union. One article celebrated the test pilot career of Marina Popovich, wife of cosmonaut Pavel Popovich. See "Marina Popovich–Test Pilot," *Soviet Life*, January 1973, 38–41. For representations of other women of the Soviet space program, see Vladimir Gubarev, "Women in Space Research," *Soviet Life*, April 1975, 36–37. Also related is an article on (male) cosmonaut's wives in the special issue celebrating the 20th anniversary of Gagarin's flight: "Perhaps the Hardest Part Is Waiting," *Soviet Life*, April 1981, 42–45.

39. Roshanna P. Sylvester, "She Orbits over the Sex Barrier: Soviet Girls and the Tereshkova Moment," in *Into the Cosmos: Space Exploration and Soviet Culture*, ed. James T. Andrews and Asif A. Siddiqi (Pittsburgh, PA: University of Pittsburgh Press, 2011), 195–212.

40. Sylvester, "She Orbits," 197; Asif. A. Siddiqi, *Challenge to Apollo: The Soviet Union and the Space Race, 1945–1974* (Washington, DC: NASA History Division, 2000), 352–53, 362, 372, 508–9, 526. Also see Jennifer Ross-Nazzal's examination of American female astronauts in this volume.

41. For American examples, see John F. Kelly, "Plan for Developing Space," *Amerika*, July 1963, 17–23; Tom Buckley, "Thomas Paine's Arena Is the Universe," *Amerika*, September 1970, 18–20; Thomas O. Paine, "Next Steps in Space," *Amerika*, September 1970, 21; "Space Station '75," *Amerika*, November 1970, 14–15; James J. Haggerty, "The Giant Harvest from Space—Today and Tomorrow," *Amerika*, February 1971, 22; Krafft A. Ehricke, "Extraterrestrial Imperative," *Amerika*, March 1973, 44–48; "Is Anybody Out There?" *Amerika*, March 1973, 49–50; "Space Artist (Robert McCall)," *Amerika*, July 1975, 26–29. For Soviet examples, see "Spaceships Today and Tomorrow, an Interview with Konstantin Feoktistov," *Soviet Life*, August 1968, 22–23; Anatoli Andanov and Gennadi Maximov, "Space Stations of the Future," *Soviet Life*, August 1969, 24–26; Sergei Petrov, "Space Travel. Its Present and Future," *Soviet Life*, October 1970, 14; "Soviet Space Exploration: Results and Prospects," *Soviet Life*, October 1974, 20–21; Dmitri Bilenkin, "The Inevitability of Outer Space," *Soviet Life*, April 1976, 6–8; Sergei Sokolov, "Exploring the Planets," *Soviet Life*, July 1976, 22–23.

42. Nixon's March 7, 1970, report endorsing a new NASA plan argued, "Our approach to space must continue to be bold—but it must also be balanced." *Amerika*, however, repeatedly printed another of Nixon's remarks emphasizing the bold approach: "We must build on the successes of the past, always reaching out for new achievements." See, for example, "Moon: Exploring the Mysteries of the Moon," *Amerika*, February 1971, 32–35; Office of the White House Press Secretary, "Statement by President Nixon on the Space Program," March 7, 1970, http://www.history.nasa.gov/SP-4211/appen-j.htm (accessed January 11, 2011).

43. In the November 1969 issue, Armstrong's famous "giant leap" quote appeared on the cover, and the words Giant Leap appeared in a giant font on p. 2. "Man Makes His Epic Journey to the Moon," *Amerika*, November 1969, front cover; "Giant Leap," *Amerika*, November 1969, 29–36.

44. See especially "Earth: Living and Working in Space," *Amerika*, February 1971, 25, 29; "Moon: Exploring the Mysteries of the Moon," 32; "Mars: Sixty-Four Million Kilometers to the Red Planet," *Amerika*, February 1971, 36.

45. Petrov, "Space Travel," 14.

46. The article quoted Gagarin's comments at his first post-*Vostok* interview that it would not be "too long before we undertake a flight . . . to the Moon" and added that Gagarin "seriously meant what he said." Golovanov, "He Wanted to Speed Up History," 28–29. Other late 1960s articles on the Moon did not mention a Soviet piloted mission. See "Queries from Readers," *Soviet Life*, April 1968, 25; Nikolai Kozyrev, "Luna: The Seventh Continent," *Soviet Life*, August 1969, 52–53. Korolev's 1963 proposal for a human voyage to the Moon in 1963 was approved in 1964. Sluggish progress and Korolev's early death in 1966 led ultimately to official denials that the Soviet Union had undertaken a human lunar program. See Asif A. Siddiqi, *Sputnik and the Soviet Space Challenge* (Gainesville: University Press of Florida, 2003), 395–408, 461–516; Asif A. Siddiqi, *The Soviet Space Race with Apollo*, (Gainesville: University Press of Florida, 2003), 697.

47. For examples of this argument, see Valentin Mikhailov, "Exploring the Moon from Baikonur and Cape Kennedy," *Soviet Life*, August 1969, 54–55; Boris Petrov, "Earth-Moon-Earth," *Soviet Life*, January 1971, 18–21.

48. "Spaceships Today and Tomorrow," 22–23; "Sergei Korolyov: Designer of Space Rockets," *Soviet Life*, August 1969, 21–22; Petrov, "Earth-Moon-Earth," 18–21; "Moon Rover," *Soviet Life*, February 1971, 18–19.

49. "Queries from Readers: First Automatic Station on the Moon," *Soviet Life*, April 1966, 55; "Exploring the Moon," *Soviet Life*, May 1966, 22–25; "First Lunar Sputnik," *Soviet Life*, August 1966, 53.

50. Ari Sternfeld, "Flying Cosmodrome," *Soviet Life*, August 1969, 22–24; Petrov, "The Space Experiment," 10–11; "Changing Ships in Orbit, Interview with Alexei Leonov," *Soviet Life*, April 1969, 48–49; Andanov and Maximov, "Space Stations," 24–26; Petrov, "Space Travel," 14; "Before the USA and the USSR Meet in Orbit, Interview with Vladimir Shatalov," *Soviet Life*, July 1975, 24.

51. Roger Launius has discussed how spaceflight narratives employ such turning points. See Roger D. Launius, "What Are Turning Points in History, and What Were They for the Space Age?" in *Societal Impact of Spaceflight*, ed. Steven J. Dick and Roger D. Launius (Washington, DC: NASA History Division, 2007), 19–40.

52. See, for example, "Pioneers Together," 30–31; "A New Frontier (Apollo 11 Moon Landing)," *Amerika*, April 1970, inside front cover; "Man's Restless Voyage," *Amerika*, January 1970, inside front cover; Ehricke, "Extraterrestrial Imperative," 44; "The First Man in Space: Yuri Gagarin," inside front cover, 1; "In Honor of Outer Space Dreamers and Research Pioneers," *Soviet Life*, August 1969, 9; "To Honor Space Pioneers," *Soviet Life*, May 1973, 42.

53. Linda Billings, Taylor E. Dark III, and Asif Siddiqi have all recently written on space advocates' discourse about "progress." See *Societal Impact of Spaceflight*, 483–99, 513–37, 555–71.

54. See, for instance, "First Stop for Men and Rocks (Lunar Receiving Lab)," *Amerika*, April 1970, 54–55; Jay Holmes, "Apollo 12: Why Go Back to the Moon?," *Amerika*, May 1970, 46; Pariente, "Apollo 15," 30; "Skylab Experiments," *Amerika*, April 1973, 14.

55. See especially "Man's First Flight into Space," 2–5; Sisokyan, "The Road to the Stars," 10; "Twenty-Five Hours in Space," *USSR*, November 1961, 13–15; Kopalin, "First Flight to the Stars," 16–19; Andrian Nikolayev and Pavel Popovich, "In the Starry Ocean," *USSR*, November 1962, 4–7; "Why Space Research? Interview with Konstantin Feoktistov," *Soviet Life*, December 1967, 41; Pavlov, "424 Hours," 12–13.

56. See especially "A 120-Day Space Experiment," *Soviet Life*, February 1965, 61; "New Stage in Exploration of Space," *Soviet Life*, January 1965, 4–11.

57. One caption for a photo of Titov and Glenn called it a "symbol of future Soviet-American cooperation." See "Welcome Gherman Titov," 6–9. See also Yuri Somov, "Spaceman's Earth-Level Orbits," *Soviet Life*, November 1967, 54–55; "Meeting on a Familiar Planet," *Soviet Life*, August 1969, 39; Ted Rukhadze, "Orbiting the USSR," *Soviet Life*, October 1969, 18–23; Ted Rukhadze, "Good Start and Soft Landing!" *Soviet Life*, November 1970, 54–57; Boris Strelnikov, "On an Earth Orbit of Friendship," *Soviet Life*, February 1976, 2–6.

58. See especially Irina Lunacharskaya, "Preparing for the First Soyuz-Apollo Docking," *Soviet Life*, April 1973, 41–43; "Soyuz-Apollo: Project for a Peaceful Planet, Interview with Konstantin Bushuyev," *Soviet Life*, December 1973, 40–42; "The Crews of the Joint Apollo-Soyuz Flight," *Soviet Life*, October 1974, 22–27; Alexei Leonov, "Soviet-American Space Rendezvous," *Soviet Life*, January 1975, 34–37; Alexei Leonov, "Challenging Space: Soviet-American Docking Experiment," *Soviet Life*, July 1975, 16–17; "Cooperation in Space, Interview with Glenn Lunney," *Soviet Life*, December 1975, 21.

59. Edward C. Ezell and Linda N. Ezell, *The Partnership: A History of the Apollo-Soyuz Test Project* (Washington, DC: NASA History Division, 1978), 37–60, 191–93; John M. Logsdon, *John F. Kennedy and the Race to the Moon* (New York: Palgrave Macmillan, 2010), 159–73; John M. Logsdon, "The Development of Space Cooperation," in *Exploring the Unknown: Selected Documents in the History of the U.S. Civilian Space Program*, vol. 2, *External Relationships*, ed. John M. Logsdon (Washington, DC: NASA History Office, 1996), 11–15; Sheehan, *The International Politics of Space*, 55–71.

60. "Here Come the Cosmonauts!" 48. See also Bowman, "Space Travelers Meet," 2–3.

61. Roald Sagdeev and Susan Eisenhower, "United States-Soviet Space Cooperation during the Cold War," NASA, http://www.nasa.gov/50th/50th_magazine/coldWarCoOp.html (accessed on January 15, 2010).

62. Jennifer Ross-Nazzal, "Détente on Earth and in Space: The Apollo-Soyuz Test Project," *OAH Magazine of History* 24, no. 3 (July 1, 2010): 29–34.

63. Konstantin Kondrashov, "Training in Houston," *Soviet Life*, July 1975, 18–19. See also Strelnikov, "On an Earth Orbit," 2–6; Leonov, "Soviet-American Space Rendezvous," 34–37; Leonov, "Challenging Space," 16–17; Georgi Isachenko, "Earth Is Our Bearing," *Soviet Life*, December 1975, 21. On the American side, see "Cosmonauts Visit USA," *Amerika*, January 1971, 48–51; "Here Come the Cosmonauts!" 48; "Apollo Soyuz Project," 21.

64. See Ronald Roy Nelson and Peter Schweizer, *The Soviet Concepts of Peace, Peaceful Coexistence, and Detente* (Lanham, MD: University Press of America, 1988), ix, 2, 4, 15; Karl Marx and Friedrich Engels, *The Communist Manifesto* (London: Penguin Classics, 2002), 241; Vladimir I. Lenin, *Collected Works*, 4th ed. (Moscow: Progress Publishers, 1964), vol. 23, 80; vol. 31, 213–14; Nikita S. Khrushchev, "On Peaceful Coexistence," *Foreign Affairs* 38, no. 1 (October 1959): 1–3; Leonid I. Brezhnev, *Selected Speeches* (New York: Pergamon Press, 1978), 8, 31, 39, 51, 130–31, 164, 225, 232.

65. On the term's history, see especially Warren Lerner, "The Historical Origins of the Soviet Doctrine of Peaceful Coexistence," *Law and Contemporary Problems* 29, no. 4 (Autumn, 1964): 865–70; Nelson and Schweizer, *The Soviet Concepts*, ix, 7; Frederick C. Barghoorn, *Soviet Foreign Propaganda* (Princeton, NJ: Princeton University Press, 1964), 88–91.

66. See USIA Office of Research, "M-279-63, The Soviet Attack On USIA," August 29, 1963, Research Memorandums 1963–1982, Box 1, Records of the U.S. Information Agency, Record Group 306, National Archives, College Park, MD (hereafter Records of the USIA); Barghoorn, *Soviet Foreign Propaganda*, 80, 87, 98–99; Nelson and Schweizer, *The Soviet Concepts*, x; Lerner, "The Historical Origins," 869; "Mirnogo sosushchestvovaniia i ideologicheskaia bor'ba," *Kommunist*, no. 16 (1959): 7.

67. USIA Office of Research, "M-160-65, Soviet Propaganda: Themes and Priorities," April 30, 1965, Research Memorandums 1963–1982, Box 5, Records of the USIA; Roger D. Markwick, "Peaceful Coexistence, Detente and Third World Struggles: The Soviet View, From Lenin to Brezhnev," *Australian Journal of International Affairs* 44, no. 2 (1990): 177; Leonid I. Brezhnev, *Leninskim kursom*, vol. 2 (Moscow: Politizdat, 1974), 275, 289–90.

68. "The First Man in Space: Yuri Gagarin," inside front cover, 1; "A Day to Remember," 2–3.

69. "Do You Know Soviet Cosmonauts?," 32–33.

70. See, for example, "Youth, a Drawing by Alexei Tertyshnikov and Igor Kravtsov," *USSR*, September 1962, front cover; "Around the Country," *USSR*, September 1962, 13.

71. Soviet propaganda and discourse routinely cited forward motion and progress. For an overview of utopian visions of the future in the early Soviet period, see Richard Stites, *Revolutionary Dreams: Utopian Vision and Experimental Life in the Russian Revolution* (Oxford: Oxford University Press, 1989), 165–223.

72. The Moral Code is printed in Grey Hodnett, ed., *Resolutions and Decisions of the Communist Party of the Soviet Union*, vol. 4, *The Khrushchev Years, 1953–1964* (Toronto: University of Toronto Press, 1974), 247–50.

73. *Soviet Life* often reprinted the Moral Code. See "New Party Rules," *USSR*, October 1961, 15; "Function of the State," *USSR*, December 1964, 4–7.

74. Yuri Gagarin, *Doroga v kosmose* (Moscow: Foreign Languages Publishing House, 1961). I discuss the correlation between the Moral Code and Gagarin's autobiography more deeply in Trevor Rockwell, "The Molding of the Rising Generation: Soviet Propaganda and the Hero-Myth of Iurii Gagarin," *Past Imperfect* 12 (2006), http://ejournals.library.ualberta.ca/index.php/pi/-article/view/1579/1105 (accessed February 9, 2008). See also Slava Gerovitch, "'New Soviet Man' Inside Machine: Human Engineering, Spacecraft Design, and the Construction of Communism," *Osiris* 22, no. 1 (2007): 135–57; Slava Gerovitch, "The Human inside a Propaganda Machine: The Public Image and Professional Identity of Soviet Cosmonauts," in *Into the Cosmos*, 77–106.

75. Gagarin, "Road to Outer Space," July 1961, 24–29; August 1961, 7–11; September 1961, 16–23.

76. "Yuri Gagarin's Own Story," 4–5.

77. Klaus Larres and Kenneth A. Osgood, ed., *The Cold War after Stalin's Death: A Missed Opportunity for Peace?* (Lanham, MD: Rowman and Littlefield, 2006), especially Lloyd Gardner, "Poisoned Apples: John Foster Dulles and the 'Peace Offensive,'" 73–93.

78. Memorandum from the Deputy Assistant Secretary of State for International Organization Affairs (Walmsley) to the Secretary of State, August 13, 1959, *FRUS, 1958–1960*, vol. 2, 169–70.

79. United States Information Agency, "USIA CA-2852, Subject: Implementation of the Five Media Priorities," March 17, 1964, Historical Collection, Subject Files 1953–2000, Box 14, Records of the USIA; Memorandum from Secretary of State Rusk and the Director of the U.S. Information Agency (Murrow) to President Kennedy, June 8, 1961, *FRUS, 1961–1963*, vol. 5, 241–42; Memorandum from the President's Special Assistant (Schlesinger) to the President's Special Assistant for National Security Affairs (Bundy), June 19, 1961, *FRUS, 1961–1963*, vol. 25, 239–40; National Security Action Memorandum No. 61, July 14, 1961, *FRUS, 1961–1963*, vol. 25, 244; Nicholas J. Cull, *The Cold War and the United States Information Agency: American Propaganda and Public Diplomacy, 1945–1989* (New York: Cambridge University Press, 2008), 199–200.

80. See Parsons, "'Faith-7,'" 12.

81. Victoria Smolkin-Rothrock has analyzed how Soviet propagandists used cosmonauts to promote scientific atheism and discussed the persistence of religion in the Soviet Union. See Victoria Smolkin-Rothrock, "Cosmic Enlightenment: Scientific Atheism and the Soviet Conquest of Space," in *Into the Cosmos*, 159–94.

82. "Our Impossible Goal," 40.

83. "*Man's Future in Space*. Painting by Robert McCall," February 1971, front cover; "Space Artist (Robert McCall)," 29; "Lunar Landing," *Amerika*, July 1975, back cover. See also Mitchell Jamieson, "Painting of Astronaut Gordon Cooper," *Amerika*, October 1967, front cover; Arthur C. Clarke, "Next—The Planets," *Amerika*, November 1969, 33.

84. Kenneth A. Osgood, *Total Cold War: Eisenhower's Secret Propaganda Battle at Home and Abroad* (Lawrence: University of Kansas, 2006). See especially chapter 8, pp. 253–87.

85. For examples, see "Close-ups of Moon Show Three-Foot Craters," *Amerika*, November 1964, 59; "Here Come the Cosmonauts!" 48–51; "Welcome Back! (Apollo 11)," 62–63; "A Day in Outer Space," 27. Photographs highlighted the many cameras and telescopes Americans used to view launches. See Stansbury, "On Target," 9; Bowman, "Space Travelers Meet," 2–3; Gregory, "As a Nation Watched," 39.

86. Gregory, "As a Nation Watched," 37; Stansbury, "John Glenn," 5; "On the Moon (Apollo 11)," 36–37. Text and images also highlighted similar scenes of public interest in spaceflight satisfied via television around the world. See especially Wilford and Wooten, "To the Moon and Back," 7.

87. "Here Come the Cosmonauts!" 48–51. For photos of Walter Cunningham and Alexei Leonov riding rocket rides at Disneyworld and Disneyland, see "Picture Parade—Astronaut's Day Off," *Amerika*, April 1968, inside front cover; "Picture Parade," *Amerika*, July 1975, inside front cover.

88. Philip Eisenberg, "A New Home in Space (Skylab)," *Amerika*, April 1973, 11.

89. See, for example, Gregory, "As a Nation Watched," 36–39; "A Day in Outer Space," 27–28; Eisenberg, "A New Home," 11.

90. See especially "On the Moon (Apollo 11)," 36–37; "Welcome Back! (Apollo 11)," 62–63.

91. See especially Gregory, "As a Nation Watched," 39; Stansbury, "John Glenn," 7; "Happy End to a Successful Space Flight," *Amerika*, November 1965, 34–35; Wilford and Wooten, "To the Moon and Back," 4; "On the Moon (Apollo 11)," 36–37.

92. "The Apollo Crew in Moscow," *Soviet Life*, April 1964, 6–9; Strelnikov, "On an Earth Orbit," 4.

93. "From a Needle to a Car," *Soviet Life*, April 1968, 41. For articles linking human space exploration with economic progress, see "Soviet Science Looks to the Future," *USSR*, August 1961, 1–2; "Communism is Coming Soon," *USSR*, September 1961, 6–13; "Lodestar of Science, an Interview with Academician Nikolai Semenov, Nobel Prize Winner," *Soviet Life*, May 1968, 36–39; Petrov, "Earth-Moon-Earth," 18–21.

94. Alexei Gorokhov, "What's for Dinner in Space?," *Soviet Life*, July 1975, 24–25.

95. Vladimir Makhotin, "Multinational Press Center," *Soviet Life*, December 1975, 22–23.

96. See especially "Progress Material and Spiritual," *Soviet Life*, November 1972, 3.

CHAPTER 7

Bringing Spaceflight Down to Earth:
Astronauts and *The* IMAX *Experience*®

Valerie Neal

A scene fills the giant screen—bare legs dangle below light blue shorts, elbows extend from dark blue polo shirts, hair billows around someone's head. Although backs are turned toward the camera, it becomes evident that we are watching a woman and two men moving about in close quarters. One drifts up toward a window as another rises out of his seat and easily floats toward her. In the next scene, six people dive into view, one at a time, twirling suspended in midair as the narrator comments, "For us inside the spacecraft, there's a new experience: weightlessness."

The Dream Is Alive, an IMAX® documentary feature film released in 1985, introduced the space shuttle and its astronauts to audiences numbering in millions around the world.[1] Produced by IMAX Corporation and presented by the Smithsonian National Air and Space Museum (NASM) and Lockheed Corporation in cooperation with the National Aeronautics and Space Administration (NASA), this film, one of the first to be shot on location in space, aimed to bring spaceflight down to Earth by faithfully capturing the astronauts' experience.[2] Edited from footage shot on three 1984 shuttle missions, *The Dream Is Alive* oriented audiences to the new spacecraft and the new generation of astronauts living and working in space. Light on narration and dialogue in favor of vivid imagery projected on a screen several stories tall, the film visually invited viewers to enter the scenes as if they were in space themselves. This immersive experience, already an IMAX hallmark, had never been attempted for spaceflight.[3]

The Dream Is Alive proved so popular that IMAX Corporation made five more documentaries in space, most recently, the Warner Bros. Pictures and IMAX Filmed Entertainment presentation of *Hubble 3D* in 2010. From first to last, these films presented new images of spacefarers, supplanting the image of astronaut as heroic pilot popularized in the 1960s. The IMAX films featured the new demographic: the many shuttle era astronauts, some of them women, who were scientists and engineers serving in the new role of mission specialist. Inspired by NASA's early theme for the shuttle age, "Going to Work in Space," these films accented competence and teamwork more than courage as astronaut crews went about their various jobs.[4] Only briefly on Skylab (1973–74) had astronauts lived and worked in orbit doing tasks that set the stage for more ambitious work on shuttle missions: scientific research, equipment repairs, satellite deployments and retrievals, and space station assembly and resupply. Shuttle astronauts were different from their predecessors; they commuted to work in space and figuratively wore hard hats or lab jackets on the job.

Audiences who "met" these new spacefarers on screen enjoyed virtual face-to-face contact and behind-the-scenes access to their training and flights. Through the vicarious IMAX experience, they could begin to understand what it was really like to be an astronaut. Over the course of six films, the astronauts became more nuanced and vocal personalities, ultimately speaking directly to the camera (hence the audience) to explain their work and share their emotional experience of spaceflight. As they demonstrated the skills and thrills of living in space, the astronauts revealed themselves as down-to-earth people who were both capable and likable. Without exit surveys, we cannot know whether viewers consciously changed their opinions about astronauts, but the films clearly offered new perspectives on the spacefaring profession. Film reviews from *The Dream Is Alive* to *Hubble 3D* indicate that the visually immersive encounters made a favorable impression (Figure 1).

For authenticity, the astronauts became cinematographers, planning and shooting footage to capture the experience of spaceflight directly, unmediated by a professional film crew. They mastered both the technical skills and aesthetic judgments of filmmaking, adding to their repertoire a new role that they reportedly enjoyed and took pride in. They had considerable latitude, in concert with the film directors, to shape their own image as spacefarers. The IMAX films thus effectively brought

FIGURE 1. Promotional art for five of the six IMAX films shot in space featured dramatic views of astronauts working outside to visually draw viewers into orbit even before they entered a theater. © 1985 Lockheed Martin Corporation and Smithsonian Institution from the IMAX® film *The Dream Is Alive*.

spaceflight down to Earth for astronauts and audiences alike, delivering a sense of the beauty, routine, and occasional drama of spacefaring as the astronauts experienced and recorded it. Risk sat in the backseat, as the motive driving these documentaries was excitement about, if not advocacy for, the shuttle program.

Shot candidly and mostly without artifice, the IMAX films reflected the production team's optimism about "routine" spaceflight. There was no questioning the aims of the shuttle era, just a pervasive confidence in a new age of exploration and discovery. *The Dream* especially channeled NASA's positive rhetoric about the benefits of the shuttle and human spaceflight in the heady years before the 1986 *Challenger* tragedy. In the later films, at least a passing acknowledgment of risk tempered this positive outlook, but the tenor of *The IMAX Experience*® remained upbeat and inspirational. The filmmaking collaboration and participating astronauts aimed to tell a good story that would inform, awe, and excite audiences about spaceflight.

The in-theater experience of spaceflight became possible through the convergence of three new technologies: the space shuttle, the IMAX 70mm film cameras, and the IMAX large-format projection and surround-sound system. The shuttle was spacious enough to accommodate the bulky cameras as a special payload, and with typically five or seven crewmembers aboard, mission planners could schedule at least one astronaut to handle shooting. The IMAX cameras produced high-resolution imagery of remarkable clarity and detail, shown to maximum advantage in advanced-design theaters. Repeated flights of the cameras yielded opportunities for a rich variety of footage, and lessons learned from one mission could be factored into the next fairly quickly. The convergence of these technologies—and the technically savvy astronauts who shot the scenes—made for a spaceflight film experience unlike any other.[5]

New York Times film critic Vincent Canby immediately recognized the impact of IMAX and other new big-screen productions. He raved that these "possibly seminal new films" were "the most viscerally exciting, mind-expanding movies being made today—the kind that provide windows on worlds previously undreamed of." Citing the Smithsonian and the American Museum of Natural History in New York, where he saw the "splendid" *The Dream Is Alive*, he reported that the film astonishes; "you are an astronaut peering at Earth below . . . you don't feel as if you're outside looking in . . . as much as the eye, the mind is engaged through sheer sensation and the multitude of details."[6]

Twenty-five years later, when *Hubble 3D* appeared, a new generation of reviewers had nearly identical responses: "The spectacular new IMAX film . . . not only puts you in space but lets you travel through it with a speed and wonder."[7] The sometimes grouchy Roger Ebert called it "awesome," and a leading critic of documentaries claimed that "what you can't anticipate until the lights dim and the film blasts off is the extraordinary exhilaration of joining astronauts on their space mission . . . The 3D puts you in the middle of the astronauts . . . and carries you in to space . . . and gives you an astronaut's-eye view of Earth."[8] A *Variety* film critic gave a stellar review: "*Hubble 3D* comes as close as any film to reproducing the curious, cosmic sensation of floating through outer space . . . it's an experience so pure and vivid."[9]

Origin of the IMAX Space Films

The Smithsonian National Air and Space Museum had barely opened its new building on the Mall in Washington, D.C., when Director and Apollo 11 astronaut Michael Collins had an idea. In November 1976 he proposed to astronaut colleague Joe Allen, who was then a senior manager across the street at NASA Headquarters, "that an IMAX motion picture camera be carried aboard and used in one of the early space shuttle flights." Collins urged that in-flight filming would "give large numbers of people views of the earth which closely simulate actually being in orbit." He reasoned that just as the immensely popular MacGillivray Freeman film *To Fly* gave people a sense of flight, footage filmed in space would give people a sense of spaceflight.[10]

IMAX Corporation, a newcomer in the film industry, had made a dramatic entrance with its revolutionary large-format camera and 70mm film technology, remarkably sharp images, gigantic projection wall or dome, and vibrant sound systems. The Canadian company developed its technology in the late 1960s and debuted it in the United States at the United States Pavilion at Expo '74 in Spokane, Washington.[11] At that time, the space shuttle was beginning to take shape in southern California. IMAX produced and distributed films outside the Hollywood market. It specialized in documentaries about thirty to forty minutes long, suitable for screening in museums, planetariums, and other educational venues. Its niche was films for world's fairs and other expos. The high resolution afforded by 70mm film and a projection screen that filled the audience's field of view gave viewers vivid imagery, a sense of motion, and a sense of being *within* the film. IMAX films were billed as "experiences" and audiences readily attested to their realism, even to the experience of motion sickness.

Graeme Ferguson, camera coinventor and cofounder of the company, became not only a director-producer of the films but also the chief salesperson for IMAX theater installations. He passionately believed that the new National Air and Space Museum, scheduled to open during the 1976 U.S. bicentennial celebration in Washington, D.C., must have an IMAX theater. Two years earlier he began to court the museum, but director Michael Collins, who had not yet seen an IMAX film, was not convinced. By coincidence Collins gave a speech at Expo '74, and an IMAX representative coaxed him to visit the United States

Pavilion, see the featured show, and examine the projector. Collins, an engineer, was smitten with the technology.[12]

Now convinced that IMAX could deliver a highly realistic experience, Collins directed that the theater for the new museum be designed for it. He also commissioned an aviation film to show off the medium and make viewers feel they were flying in a variety of aircraft. When the Museum opened to the public, *To Fly* vied with historic aircraft and spacecraft as the most impressive visitor experience. The museum's IMAX theater was only the sixth one built in the United States, and it immediately became a tourism magnet in the nation's capital.[13] The mutually beneficial relationship between IMAX Corporation and the museum lasted more than two decades; all six space films premiered there.

Taking his cue from Collins, Ferguson lobbied NASA to fly an IMAX camera on the shuttle. He argued that "an IMAX film about the Space Shuttle . . . would provide viewers for the first time with a realistic sense of what it is like to be in space. Such a film would have great educational value, and would also assist NASA in informing the public about its work."[14] NASA initially responded positively to the proposal because such a film "would receive broad exposure, present the STS [Space Transportation System] in a favorable light, [and] allow the public to 'participate' on a Shuttle flight."[15] But approval did not come until several years later, when the shuttle's test flights ended well and the vehicle was declared operational. NASA Administrator James Beggs made the decision to fly the IMAX camera after seeing *Hail Columbia!* A number of NASA executives continued to provide crucial support over the years for production of the space film series.[16]

The first IMAX film to feature the space shuttle, *Hail Columbia!* (1982), was not created for the National Air and Space Museum but enjoyed a long and popular run there. A dramatic documentary about the first space shuttle mission, *Hail Columbia!* featured IMAX sequences filmed on the ground, culminating in a thunderous, body-rattling launch shown repeatedly from various angles for maximum effect. No sponsor emerged to underwrite production costs, but Ferguson believed that this signal event in U.S. spaceflight was so important that IMAX Corporation filmed it independently. *Hail Columbia!* premiered in the museum for NASA's twenty-fifth anniversary in 1983. The film became a sensation.[17]

Collins's request to NASA and Ferguson's persistence in cultivating support for the idea ultimately led to a collaboration between the Smithsonian, IMAX, NASA, and corporate sponsor Lockheed (later Lockheed Martin), a major NASA contractor for the shuttle program. Together they created three documentary films made in space. Beyond the first three films, some combination of the collaborators created three more IMAX space films.[18] Vigilant about accuracy and its public image, NASA had always preferred to make its own films about space exploration and declined to endorse commercially made ones. However, anchoring the IMAX production in the well-respected Smithsonian, treating the films as documentaries, and seeing one of its own contractors fund the films as a public service gave NASA the confidence to make this collaboration possible. All members of the project team had a stake in presenting the shuttle and the astronauts in a favorable light.[19]

The Dream Begins

By the time *Hail Columbia!* opened in the museum's theater, the group was well along in planning a successor to be filmed in space by shuttle crews using an IMAX camera. The new director of the museum, aviator and writer Walter Boyne, was very enthusiastic about the project. Together with Graeme Ferguson, Boyne supplied creative energy for *The Dream Is Alive*.[20] While Ferguson worked out the technical logistics of the project, Boyne tackled the logistics of working together.

For several years, Ferguson had been meeting with astronauts and payload managers to determine how the large camera could best be accommodated and how much crew training would be needed. He obtained good advice and encouragement from many.[21] A designated patron at NASA Headquarters, William Green, guided him through the bureaucracy and worked closely with him to identify missions that offered the best opportunities for scenes that would flow together as a coherent story. By 1982, IMAX had worked out most of the logistical issues with NASA and outlined a film based on three 1984 missions. One was a two-satellite delivery mission, another was a satellite retrieval and repair mission with spacewalks, and the third was a three-satellite delivery.[22]

Lacking a precedent, the group had to work out guidelines for their respective roles. IMAX would have creative control, the museum would be responsible for historical and technical accuracy and have final say on content, and the museum and Lockheed would share review and approval rights. NASA would provide the flights and technical support but had no formal role in developing or approving the creative or factual content. All were committed to the goal of making a first-class production that would be accurate, authentic, enjoyable, and of the highest quality.[23]

The museum drafted a set of general principles for the film to have appropriate elements of "visual splendor, scenic beauty, humor, visceral excitement, . . . human touches, excellent music, and so on . . ." and stated that "the film is explicitly intended *not* to be a promotional piece for NASM, NASA, Lockheed, IMAX, or anyone else, but is rather to be an uplifting emotional and educative experience for people everywhere." The group agreed that the film would be "above all an artistic endeavor" dependent on "the artistry, inventiveness, innovation, and ingenuity of the film maker . . . to optimize public enjoyment and the success of the film."[24]

As the group worked out the relationship mechanics, NASA and IMAX had to find a way to place the IMAX equipment on board the space shuttle. Typically, every available inch and pound were spoken for, so determining how to accommodate the camera, lenses, film magazines, and lighting and sound equipment was a challenge. Engineering and crew assessments validated the 80 pound camera's use in orbit; it would fit into a middeck stowage locker and could be handled within the crew cabin. Mounts were designed to position it at the aft starboard window to shoot activity in the payload bay or in an overhead window to shoot views of Earth. Flight tests in NASA's weightlessness training aircraft indicated that astronauts should be able to hold and use the camera without difficulty in space. The IMAX camera could meet the primary goal of the project: capturing the astronauts' point of view in order to bring spaceflight down to Earth.[25]

In 1983, NASA granted approval to fly the camera, train the crews, and make the film. IMAX had no sooner received permission to fly the in-cabin camera on the 1984 shuttle missions than the project team began thinking more boldly. Could a camera be installed in the payload bay for a better view outside the window? Could a camera be attached to the shuttle's robotic arm and aimed anywhere? Could a camera be mounted on a companion free-flyer satellite to shoot the shuttle from a distance? Taking the camera outside was technically more challenging than shooting through the windows. There simply was not enough time to develop, test, and put on board a pressurized container for the camera and figure out how to operate it remotely for the first three flights, so the idea of an extravehicular camera was kept alive as a future goal.[26]

Astronauts Become Cinematographers

It was IMAX Corporation's responsibility to train the crews to operate the camera, and its technical team learned that the astronauts, most of them engineers, were quick studies. A cadre of astronauts took a keen interest in the camera itself and in the filmmaking project. They realized that IMAX technology could indeed bring the experience of spaceflight to the public, and they were eager to participate.[27]

The crews' challenge was not only to learn how to operate the complicated camera, change film, and do other routine tasks but also to learn how to set up shots at the right exposures and use the right lenses and lighting to get the best possible images. The astronauts needed to become more than adept photographers; they also had to become accomplished cinematographers. IMAX and

NASA worked together to develop the crew training schedule and curriculum. In sessions totaling about thirty-three hours over six to seven months, each mission crew underwent training first in the mechanics—learning how to use the camera and choose lenses, set exposures, focus properly, work with lighting and sound, and understand various principles specific to the large format.[28]

The curriculum's second part focused on aesthetics and scene development under the tutelage of film producer-director Ferguson and writer-editor Toni Myers. The astronauts watched IMAX films as their instructors gave running commentary on composition, editing, narrative flow, and the dramatic arc of a story. After this familiarization the crew carried out assignments to conceive, set up, and shoot a scene in the crew compartment trainer that they critiqued in the next training session. These exercises in scene development and self-review polished their mechanical skills and cultivated their eye for the aesthetic elements of filming.[29]

It was crucial for the astronauts to gain proficiency before the mission because the film rolls taken into space allowed for exactly three or eight minutes, parceled into thirty-second shots. Every scene would be shot in one take, with no repeats, so every second of film must count. Proficiency meant being individually skillful and responsible for aspects of the shoot normally performed by a professional production team, with the additional requirement to make no mistakes. Although missing a shot would not have been fatal, men and women already committed to the zero-tolerance-for-errors work ethic of spaceflight proved equally vigilant not to waste one second of film or opportunity. They also proved clever at problem solving to ensure success.[30]

Any task or payload added to a mission must compete for the crew's limited time in space and in their intense preflight training schedule. Because Ferguson had spent years working within the appropriate NASA organizations and because there was both broad interest in making the film and a management directive to support the effort, the project encountered no overt resistance and had some very strong advocates, including the director of flight crew operations, George Abbey, who managed all astronaut assignments. The burden fell on IMAX to define the impact on crew time, develop clear requirements and rationales for training, and define the parameters for all desired shots. NASA occasionally disapproved an IMAX activity that interfered with a higher priority but generally strove to accommodate the filmmaking.[31]

By 2010 when *Hubble 3D* appeared, 145 astronauts and cosmonauts (about half the shuttle era corps) had trained to use the cameras and participated on IMAX missions. They were crewmembers on 24 flights selected on the basis of mission plans that could yield a compelling film scenario. Some flew with the cameras more than once and grew quite adept as filmmakers. Astronaut

Marsha Ivins, who became heavily involved in the IMAX projects and likely reflected the attitude of others, remarked that it was fun to do the filming and the crews were proud of their work. Seeing the films on the big screen and knowing they had been part of showing people what space was like was "almost as good as saying I went to space."[32]

Plotting to Capture the Astronaut Experience

The goal of *The Dream Is Alive* and the later films was to use the techniques and artistry of filming in space to capture the spaceflight experience. With astronauts as cinematographers, audiences would see and hear what happened in the moment. There would be no special effects of sunrise or weightlessness or spacecraft motion; it would all be real. Yet the film would not be random scenes in space. It would have the structure and flow of a good story.

A directorial vision and planning informed the project. IMAX had already developed storyboards—the visual and content map—of the film before crew training began. In artist's illustrations, photographs, and skeletal script, these pages presented a story of the space shuttle in service. It was not an imaginary story but an anticipatory one, conceived by close study of mission plans. The concept for *The Dream* opened with scenes of *Challenger* returning from orbit as *Discovery* was being prepared for launch. Script notations conveyed the essence of *The* IMAX *Experience*®: "You are in the pilot's seat" as the shuttle hurtles into space or the runway appears to greet you. "We can see the Earth as the astronauts see it," moving high above clouds, sea, and fascinating landforms toward an ever-receding horizon. "We feel we are actually there" in scenes of the crew in weightlessness. On one of the missions, the crew would retrieve, repair, and redeploy the Solar Max orbital observatory. The proposed film would feature their training and flight, culminating in the orbital pas de deux of the large shuttle and smaller satellite and of the two spacewalking astronauts. The storyboards suggested a vision of future space stations and the beginning of a new age of exploration. So well had the IMAX team and shuttle crews done their homework that the actual film varied only slightly from this concept.[33]

By early 1984 IMAX produced a more detailed concept for the viewer's experience. Complementing the pictorial storyboards, this scenario outlined the story in more detail and described the filmmaker's intentions for music, sound and visual effects, and other elements of style. Already the film team could hear the "musical pulse" of a not yet composed score conveying an "atmosphere of energy and activity" around the idea of an operational shuttle fleet and "the adventure, the humor, the hazards of life in zero-gravity." The mood of the film would range from lively to taut to triumphant as narration and music carried the visual story—and its inherent optimism—forward.[34] The team agreed that renowned CBS News anchor Walter Cronkite should narrate *The Dream Is Alive*. He readily consented almost a year before filming began.[35]

Working with the film outline and narrative, IMAX and NASA technical staff pored over the timelines for the three selected missions. They drew up shot lists of desirable scenes for the story and then matched the shots to a host of other parameters, such as orbiter ground track, time of day/night, Sun angle, to build a shooting schedule within the crews' master schedules. Timing was of the essence; sometimes there would be only seconds to shoot the desired scene as the orbit moved over the Earth or the Sun angle was just right, so equipment preparation time also had to be calculated and scheduled down to the hour, minute, and second. The time-limited onboard film and the pressure to get each scene in only one take required this detailed preplanning. The scenes that ultimately flowed together seamlessly in the final film as if captured fortuitously were precisely calculated. The shot list also allowed for opportunistic scenes at the crew's discretion, and they responded with surprises that delighted Ferguson and Myers.

Filmed in Space by the Astronauts

The 1984 shuttle missions approved for IMAX filming were packed with action and spacewalks, so they held good promise for interesting perspectives. They also were fine examples of the shuttle's capability as delivery truck and service station, NASA's predominant message about the shuttle program at the time. IMAX planned to do the in-orbit filming in fairly quick succession and then release the film in mid-1985.

The crew of the first selected mission (STS 41-C in April) had two main tasks: to deliver a science and engineering research satellite called the Long Duration Exposure Facility and to capture and repair the ailing Solar Max observatory. This satellite had not been designed for in-orbit servicing, but an attempted repair was a good opportunity to try tools and techniques that would soon be needed to keep the Hubble Space Telescope in good working order. The intended dramatic high point of the film would be an astronaut's untethered free flight in a propulsion backpack, the Manned Maneuvering Unit, to grapple Solar Max and bring it to the payload bay. The mission became more dramatic when the planned capture method failed, and the crew and mission support team had to develop a new approach. *The Dream* recorded the suspense and tension of a mission threatening to go awry but saved by ingenuity.

The second selected mission (STS 41-D) started with a first, a dramatic main engine shutdown on the pad, during its June launch attempt. This mission was scheduled to deploy three communications satellites and a large solar array experiment. As the technical problem that caused the shutdown was being resolved, NASA cancelled another mission, combined the payloads, and rescheduled 41-D for the end of August. This reconfiguration and launch delay put the film two months off schedule and forced IMAX to revise the production plan. Shifting launch dates could affect the timing and lighting of planned shots and the seasonal appearance of natural features on Earth. It became evident that filmmaking in space called for some flexibility.

The third selected mission (STS 41-G in October) was primarily a science and engineering mission to evaluate techniques for doing complicated tasks in space. Doing so, Kathryn Sullivan became the first American woman to do a spacewalk, which was captured by the IMAX camera. This mission also represented NASA's claim that human spaceflight was becoming routine. Commander Robert Crippen already was making his fourth shuttle flight, and Sally Ride was on board for her second flight. Tasks included operating an imaging radar system and other Earth-observing instruments and conducting an orbital refueling task to rehearse procedures for an upcoming mission.

For each mission, the commander appointed one crewmember to be in charge of the IMAX camera. On the Solar Max mission and the third mission, Commander Robert Crippen put himself on IMAX duty; pilot Michael Coats did the honors for the second flight. Typically, all crewmembers became involved in the shoots, whether as subjects, lighting and sound technicians, camera operator, or onboard director.

The Dream project team was pleased to see how the fourteen astronauts who handled the camera made the film their own.[36] Calling it "Max," they treated it almost as a crewmember and remained attentive to its schedule. Although they worked from preplanned shot lists and storyboards, the crews seized moments for unanticipated shots. Film director Graeme Ferguson recalled that some of the best scenes were impromptu, such as an eerie scene of astronauts asleep. He found them to be so accomplished at filming that there was no reason to consider flying an IMAX specialist aboard the shuttle (Figure 2).[37]

While the missions were under way, Ferguson and the IMAX team occupied a station in NASA's Customer Support Room in the Mission Control Center in Houston. From their console they watched onboard television, listened to the flight crew's conversations among themselves and with Mission Control, and consulted with other mission support engineering teams. The IMAX ground team often could monitor the camera in the shuttle so well that they knew every time and for how long it operated and could judge how much film was exposed or remaining. Through the astronaut on duty as CAPCOM (capsule communicator) in Mission Control, the onboard astronauts and IMAX ground team consulted on such creative matters as how to set up shots in a mutually beneficial interaction.

FIGURE 2. The bulky in-cabin IMAX camera mounted to flight deck windows for stability. Pilot John Blaha is filming spectacular views of the Earth from *Discovery* on the STS-29 mission in 1989. Courtesy of NASA.

After the mission, the team screened the rough footage, or "rushes," with the astronauts. On the basis of what they saw from one mission, they made adjustments to the planned shots and techniques for the next mission. After the three 1984 flights, all major events and most of the specified scenes had been covered. The returned film magazines headed to the processing lab and on to production into a show that appeared on screen at last, nine years after the initial idea. The signature credit line on the first and later films announced, "Filmed in Space by the Astronauts."

The credits also listed each mission and every crewmember involved in the film, whether or not that astronaut had appeared on screen. In working with the spacefaring crews, Toni Myers, who wrote and edited all the IMAX space films and directed or produced the later ones, grew comfortable telling the astronauts, "Remember, you're the directors, not me. If something interesting comes up . . . go for it." She did not want them to be too constrained by storyboards and shot lists, and she admitted to being amazed at the quality and ingenuity of their work. The astronauts enjoyed capturing their experiences to share with others, and they made many of the decisions about what to film.[38] Their longtime training manager James Neihouse said, "They do such a great job and are so proud of their film work. It's their film. We're just the facilitators."[39]

Astronauts in IMAX Films

Astronauts on missions with IMAX cameras found themselves in an unaccustomed role: film stars. Not only were they shooting the scenes in the shot plan, they also were documenting their own activities in space. On camera, they were to be themselves, not actors, and to carry out their work naturally as if the camera were not present. Scene setups were not contrived staging; they were efforts to capture mission activity as it happened, with interference only to get the best exposure and focus in the most advantageous light. At the same time, the crews played a supporting role to the true star of the films, the footage itself—large-format, high-resolution, well-framed scenes that would fill a huge screen.

The series of IMAX films from *The Dream Is Alive* to *Hubble 3D* spanned twenty-five years of spaceflight and featured astronauts who joined NASA over five decades. Although the films had similar narrative and visual elements, they had no formula for the image of spacefarers. The shuttle era astronaut corps was anything but homogeneous, and fewer matched the fighter-pilot template for astronauts of the previous era.[40] Arguably the most direct encounter millions of people had with spacefarers, the IMAX documentaries added new dimensions to the iconic image. The first film introduced the public to shuttle astronauts, and the continuing IMAX experience enriched the astronaut image.

Three aspects of the IMAX space films reveal how that enrichment occurred: through scenes of astronauts as "ordinary" people at work, increasing candor in the individuals' voices and interactions with the camera, and eventual recognition of risk. Influenced by events on the ground, the crews' comfort on camera, the two shuttle tragedies, and the IMAX team's deepening respect and fondness for the astronauts, images of spacefarers become steadily more multidimensional and personable in successive films. Traits associated with astronauts emerged more by nuance through their unstudied humanity than by direct narrative statement.

Image: Heroic Figures or Human Beings

Hail Columbia! (1982) had featured the first shuttle crew, astronauts John Young and Robert Crippen, with some of the awe expressed for the first generation of astronauts. Dean of the astronaut corps, Young was already a four-time space veteran with two flights to the Moon and a top-notch test pilot. He and Crippen in their flight suits had the aura of military pilots and as the first to fly an untested spaceship deserved the mantle of courage. The film dwelled on the pilots as they appeared in two settings: with their T-38 jets and the shuttle and meeting the press before and after their mission in space. With the vehicles they were all business; with the press they were sardonic, joking, and eloquent by turns. With just enough swagger, they brought to mind the heroic astronaut archetype of the 1960s and of Tom Wolfe's novelistic history of the first astronauts, *The Right Stuff*, released as a feature film a few months after *Hail Columbia!* The dramatic tension of the first shuttle flight, plus President Ronald Reagan's sendoff prayer and a postmission ticker tape parade, fortified the classic image of astronauts as heroic pilots.

Except for the mission commanders, the astronauts in *The Dream Is Alive* were new recruits for the space shuttle era. Selected in 1978 and 1980, they were at least ten years younger than those already in the corps, and most of them arrived with quite different backgrounds. The majority were "mission specialists," graduate-degree engineers and scientists responsible for much of the work in orbit except flying the spacecraft. Commanders Crippen and Hartsfield were in the last class of the preshuttle era, a group that had transferred to NASA from a cancelled Air Force space program in 1969. They came from the military fighter pilot/test pilot tradition, as did newly selected shuttle pilots.[41]

The first two astronaut classes of the shuttle era gained attention for their diversity. For the first time, women, African Americans, an Asian American, and a Hispanic passed muster for selection. In accord with NASA's own deliberate policy and changes in American society reflected in and required by law, the astronaut corps was no longer segregated by sex or color. By the seventh shuttle mission, launched in 1983, the "new guys" were going into space. The astronaut corps had changed, and the IMAX films played a seminal role in introducing the new spacefarers.

The admission of women astronauts stirred considerable gender concerns inside and outside NASA.[42] When the women were introduced to the press, the questions were predictable. Had NASA relaxed the requirements in order to select women? Would women be physically able to do the job, or would they get special treatment? How would they fit into a program steeped in male experiences and traditions? How would privacy concerns be addressed in the close quarters of a spacecraft? Despite the opportunities spurred by social change in the 1960s and 1970s, some skepticism greeted the first women astronauts.[43]

By the time of the 1984 missions, most of the unease had passed. Sally Ride had become the first American woman in space in a crew with four men, and Guy Bluford had become the first African American in space; both had acquitted themselves well on their 1983 missions. The number of women in the astronaut corps had almost doubled from six in 1978 to eleven by 1984, so it was clear that females were a growing presence. They worked hard to be perceived as astronauts, not *women astronauts*, to prove that the job was both sexless and genderless. The iconic image of the white male astronaut had been challenged. Yet much of the public had only vague awareness of shuttle astronauts and missions, which already seemed routine enough not to attract close media attention.

It was in this context that *The Dream Is Alive* matter-of-factly presented the unisex in-orbit life of astronauts. They had the same wardrobe: polo shirts and soft-cotton jeans trimmed in Velcro. They were equally likely to be taking photos or operating the robotic arm, preparing a meal, doing an experiment, or suiting up for a spacewalk. They shared a living space hardly larger than a group tent, as if they were friends on a camping trip. Their only modicum of privacy was in the small shared toilet alcove, so close to everything else in the cabin that it barely sufficed for modesty.

The scene described as this essay opened appeared about nine minutes into the thirty-seven-minute-long *The Dream*, after a shuttle landing and launch, after scenes of some of the thousands of shuttle workers, after views of the Earth from space, and after deployment of a large experiment. The astronauts appeared last, almost anticlimactically, in this introduction to shuttle-style spaceflight. The billowing hair, bare legs, and central position in the frame drew the eye to the unnamed female astronaut (Judith Resnik) with perhaps a momentary surprise. The scene immediately flowed into another with two women and four men, and any surprise yielded to a sense of normalcy—why would there not be women *and* men in space?

While crewmembers seemed as comfortable together as family members, the narrator explained that the crew cabin was their living room, dining room, bedroom, workshop, and study. Within one minute the film dispensed with any anxiety about men and women in space together. In another scene a shirtless man (Steve Hawley) exercised on the space treadmill while two other men (Mike Mullane and Charlie Walker) worked on experiments just inches away. Then the crew was eating together in a circle as relaxed as on a picnic, playing with their food in weightlessness. The scene of the crew asleep showed each chastely zipped into a sleeping bag strapped to the walls of the cabin. The ordinary routines of life seemed little different in space.

The film also highlighted the astronauts' work—deploying a communications satellite and an impressive solar array, flying the Manned Maneuvering Unit, retrieving and repairing the Solar Max satellite—and some of their training. As the film offered a guided tour of the shuttle and a survey of astronauts' onboard activities, it was easy for viewers to feel as if they were floating beside the astronauts and to understand that "we *can* work in space" (narrator Cronkite's emphasis). The film kept its focus on the stunning visuals as he intoned, "At last you can experience space as the astronauts do."

The Dream Is Alive had a respectful tone without overt hero worship. Although Cronkite's voice as narrator, so familiar from television in the Apollo era, implied a continuing story of heroic human spaceflight, his message, written in the preaccident glow of the early shuttle period, actually was quite different:

"already people like you and me are beginning to travel into space" suggested that spaceflight was becoming routine enough for "ordinary people" to fly. Audiences may well have considered these astronauts to be heroes, especially after the *Challenger* tragedy only months later, but *The Dream* did not directly portray them as such. The narration for the climactic effort to grapple Solar Max out of orbit accented teamwork and competence, not heroics. The astronauts' goal in filming themselves was to capture their shared experience matter-of-factly and to illustrate the message that seemingly ordinary people were now in space. Apart from a passing comment that thousands apply but few make it, the job and life of an astronaut appeared rather normal, albeit in an extraordinary setting.

Hubble 3D, filmed in part and written after the 2003 *Columbia* tragedy and toward the end of the shuttle program, had a more reflective, even valedictory script. This narration explicitly lauded the astronauts in one of the few emphatically editorial passages in the IMAX series. "In the last two decades thirty-two astronauts have put their own lives on the line to give life to Hubble. Each one of these men and women is a true hero." Praising them again, narrator Leonardo DiCaprio observed that "in future journeys we'll need all the amazing skills and teamwork of this crew, the same courage and inventiveness that has restored Hubble to its full capacity and beyond. They have exceeded every expectation." Without a shuttle to fly the cameras, this was perhaps the last IMAX film shot in space. Ferguson, Myers, and the IMAX production team who had invested a good portion of their own careers working with astronauts evidently wanted to seal the legacy of shuttle era spacefarers with the heroic imprimatur. Only the last IMAX film offered such an overt heroic tribute.

Identity and Personality

In the course of the IMAX films, the astronauts became more interesting and personable by virtue of the ways they were identified and interacted with the camera. By comparison to the astronauts in *Hubble 3D*, those in *The Dream* seemed relatively restrained, self-conscious, and almost anonymous, as if each represented the "typical shuttle astronaut." The first film had less self-revelation and improvisation than the last one in which each crewmember's distinct personality lit the screen. Although the narrators always told the audience "we are there," the relationship between astronaut and audience ranged from reserve to intimacy.

What made the difference? The manner of naming and speaking changed noticeably. In *The Dream*, astronauts were named almost as an afterthought, with only a brief mention of what each was doing. They went about their work and routines typically ignoring the camera and speaking to one another only in clipped phrases. The narrator told the audience what was happening. These shuttle crews seemed to be rather generic and interchangeable. Despite the audience's illusion of being in space with them, in comparison to later films this one treated viewers more as observers than participants and astronauts more as a group than as individuals.

The later films relied more on the astronauts' own words, first as voice-over and ultimately on camera. The second film, *Blue Planet* (1990), incorporated in audio two named but unseen astronauts musing about the awe of viewing Earth from space.[44] So many astronauts filmed scenes for *Destiny In Space* (1994) that only a few could be identified in the narration, but all 50 names appeared in the credits. In *Mission To Mir* (1995), several identifiable astronauts' voices, especially Shannon Lucid's, supplemented the narrator's with commentary throughout the film. Their unscripted conversational style added content and personality that heightened the viewers' sense of proximity, even though the speakers were off camera.

This trend toward individuality culminated in the last two films for a more in-depth astronaut experience. *Space Station 3D* (2002) more liberally used astronauts' commentary in lieu of narration. Their remarks became more introspective as they described their feelings and sense of meaning in space-flight. Finally, *Hubble 3D* (2010) ventured into a more intimate interview-style encounter.[45] Crew members had individual time on camera with their name and title shown on screen. Instead of offering disembodied voice-over comments, they looked directly at the camera (audience) and talked about what they did and how they felt about it. In long passages without a narrator's intrusion, they asked each other questions, became pensive or teasing, and spoke frankly about their emotions. They were eloquently and comfortably "down to earth."

Toni Myers always encouraged the astronauts to be themselves because she believed that in them "we see ourselves."[46] In the last film, as they revealed their distinct and engaging personalities, the dream that ordinary people can go into space still seemed alive. Although the narrator called them heroes, they really seemed more "like you and me"—interesting and lively human beings, not iconic figures. Yet that same film ended the series in a fusion of individualism and heroism. Ultimately, the IMAX films honored the trope of astronaut as hero.

The Reality of Risk

The Dream Is Alive opened with a shuttle landing, featured two launches, and ended with astronauts working outside high above Earth. Nowhere did the film explicitly acknowledge the risk inherent in riding a powerful rocket into space, living and working inside a complex but fragile spacecraft, or shooting back to Earth inside a fireball. Instead, the film conveyed the unbridled enthusiasm for routine spaceflight that pervaded the aerospace community in the early 1980s.

Like *Hail Columbia!* and indeed like NASA and its aerospace team, *The Dream* had a buoyant spirit of confidence, if not naïveté.

Any illusions about safely routine spaceflight evaporated in an instant on January 28, 1986, when the tenth *Challenger* mission went horrifically awry just seconds into ascent. *The Dream Is Alive* had been showing in theaters for several months, and on the date of the *Challenger* crew's quite public death another premiere of the film was scheduled in a new venue. Two members of the *Challenger* crew, Dick Scobee and Judy Resnik, had done much of the filming for *The Dream* on their 1984 missions, and Resnik was visible in the film. IMAX Corporation and the theaters faced a quandary: cancel showing the film or edit the scenes and credits? Instead of shrinking from the tragedy, theaters inserted a slide before each screening that dedicated it to the memory of the *Challenger* astronauts. In the following months, *The Dream* continued to draw audiences in a more reflective mood. In retrospect, *The Dream* seemed so innocent that one of NASA's senior managers noted ruefully, "it would have been impossible to make a film about space which was so light and joyous and unself-conscious *after* the accident. Imagine the pressures from all of us and on all of us to make it serious, reverent, to make it 'sell' the program." Nevertheless, in the immediate aftermath of the *Challenger* loss, *The Dream Is Alive* did "sell" human spaceflight on the shuttle; the film "is the only thing out there right now that keeps hammering away every day saying to millions of people that this program is right and wonderful and inevitable."[47]

Had *The Dream* been filmed after the 1986 *Challenger* tragedy, at least one scene would very likely have been different. Both astronauts and scriptwriter avoided the obvious in filming Kathy Sullivan and David Leestma suiting up together for a spacewalk—one of the most dangerous mission tasks. As both appeared in their long johns liquid cooling undergarments with Dave's urine collection belt visible and Kathy's padded rear suggesting the diaper she was likely wearing, there were no words about risk or courage. Instead, narrator Cronkite commented that "already people like you and me are beginning to travel into space" and soon will inhabit a space station. He went on to predict that our children may live in space and their children may be born there. An audience might be forgiven for naively seeing only two people almost mundanely dressing for work, not heroes braving the dangers of space. Might some viewers have felt anxiety about the unspoken subtext—the risk of death in spaceflight?

This silence was not broken until *Space Station 3D* (2002) addressed risk, not once, but repeatedly. In the opening sequence, viewers were startled to see a spacewalk go wrong as an untethered astronaut drifted away from an extravehicular activity (EVA) worksite. While narrator Tom Cruise spoke about the risks of working in space, the scene shifted to a virtual reality lab to reveal that the emergency was a training simulation, one of many designed to prepare crews to deal with those hazards. Risk came up again as spacewalkers tested their emergency rescue device and again in the narrator's closing comments that expanding knowledge in space is "worth all the risks"—a line that became tragically meaningful when *Columbia*'s final crew died coming home from their science mission while this film was showing in theaters. Since the 1986 *Challenger* accident, the inherent danger of spaceflight could not be ignored, but it was barely hinted at in IMAX films until fifteen years later.

Hubble 3D also made risk a main theme by opening with the line, "Seven brave astronauts are about to embark on the most challenging and risky mission ever flown in space."[48] The narration and visuals accented the risk to the telescope and to the astronauts in the complicated EVA tasks, building suspense until their celebration of a difficult mission accomplished. As noted earlier, this film went a further step to *tell* the audience that these astronauts were heroes, a reminder that spaceflight was not, in fact, routine. Previous IMAX films had not alluded to the *Challenger* tragedy, but this one mentioned the *Columbia* loss and showed a very brief clip of the debris meteor shower. Ending the series on a heroic note made a subtle salute to the deceased spacefarers as well as the Hubble crews.

As the Astronauts See It

The initial impetus for the IMAX films had been to capture spaceflight and views of Earth as the astronauts see it. Although the film projects adhered to the goal of portraying astronaut experiences as they really were, they also gradually introduced new techniques that exceeded the astronauts' direct experience. After the first set of IMAX flights, NASA granted permission to add a camera in the shuttle's payload bay to record the astronauts' spacewalking experience. Shooting outside the confines of the crew cabin permitted wide-angle views of Earth and a sense of floating free in space with "you are there" fidelity. Extravehicular activities tend to be dramatic episodes within a mission, and the payload bay camera bore witness to the thrill and the physicality of spacewalking as if seen through the astronauts' eyes (Figure 3). Mounting the exterior camera on a small retrievable satellite, however, yielded a panorama of the entire orbiter against Earth and space, a view that the astronauts could not see on their own.

As the art of computer-generated special effects and visualization advanced, the space film producers ventured from their initial documentary style toward a format with more artifice. This kept the films on the frontier of theatrical novelty and added to their thrill factor but also moved them away from a fully realistic presentation of spaceflight as the astronauts saw and experienced it. In a striking departure, *Destiny In Space* (1994) introduced scientifically credible

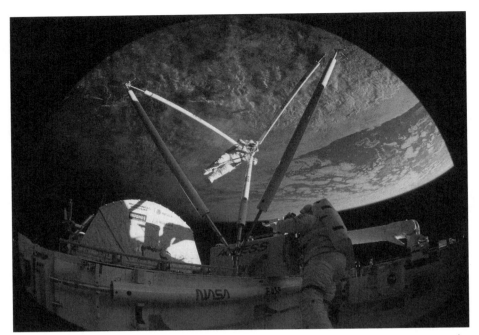

FIGURE 3. Installation of a remotely controlled IMAX camera in the shuttle's payload bay permitted wide-angle views of the Earth and astronauts at work, like this scene from its first flight on the STS 61-B mission in 1985. © 1990 Lockheed Martin Corporation and Smithsonian Institution from the IMAX® film *Blue Planet*.

computer-generated flights over other planetary terrains, well beyond the experience of contemporary crews. *Hubble 3D* (2010) ventured vastly farther from the astronauts' point of view and experience in a cosmic journey via supercomputer manipulation of astronomical images.

Space Station 3D (2002) introduced the 3-D viewing experience in a space film.[49] Although it was not obvious to viewers, the filmmaking crew spent more time planning and rehearsing their shots than previous crews, in part to adjust to the requirements of 3-D. They deliberately choreographed certain actions for the 3-D illusion, floating items toward the camera or setting up action in different depths of field. Although IMAX had always promoted the films as real experience, not artifice, the camera technology itself made some rehearsal advisable. Besides the special needs of 3-D, some complex shots required staging and rehearsal simply to ensure they could be captured correctly in one take. On rare occasions, a fairly trivial but credible action was faked; for example, the IMAX camera was so loud that no one could really sleep while being filmed (Figure 4).[50]

Astronauts and *The* IMAX *Experience*®

The Dream Is Alive reigned for many years as the most popular of the IMAX space films, attracting 48.6 million theater viewers by 2012.[51] It effectively

FIGURE 4. Astronauts used an even larger 3-D IMAX camera on the International Space Station to document their activity inside and outside the orbital research center. Astronaut Bill Shepard is checking a shot on the monitor attached to the camera. Courtesy of NASA.

introduced the space shuttle, the new generation of astronauts, and the realities of contemporary human spaceflight to the world. It also gave viewers the "Aha!" experience of spaceflight, almost as if they were astronauts themselves. Each later film extended that experience in some new dimension, by focusing on the views of Earth from space (*Blue Planet*), scientific research and planetary exploration (*Destiny In Space*), life on a Russian space station (*Mission To Mir*), the International Space Station (*Space Station 3D*), and cosmic exploration (*Hubble 3D*).

These six productions had a first-class luster that elevated the astronauts' film work to an award-winning professional level. Shown primarily in museums and similar education venues rather than commercial theaters, the IMAX films had widely successful box office appeal after their premieres at the Smithsonian National Air and Space Museum. All were well reviewed for both the extraordinary scenery and the immersive experience of spaceflight.[52] *The Dream Is Alive* fulfilled expectations for delivering to audiences the astronauts' experience of living and working in space. The later films enhanced that experience and the image of spacefarers by capturing all kinds of activities in space and encouraging the astronauts to reflect on how it felt and what it meant to be in space.

As a set, these films presented a remarkable range of astronaut experiences. Audiences had behind-the-scenes glimpses of activities and facilities they were unlikely ever to witness directly: zero-gravity training and underwater simulation, launch control and mission control rooms, Star City and the Baikonur Cosmodrome, the suit-up room and launch tower white room, the astronauts' many workplaces. Viewers saw them working hard, relaxing, being reflective, and being funny. Each film vividly showed what the job of astronaut entailed, satisfying curiosity with familiarity by opening the spaceflight experience to all.

The IMAX films also revealed more about the astronauts (and cosmonauts) than simply how they lived and worked. The films presented the new norm of men and women, Americans and other nationalities, working together companionably in space. Interesting as their daily routines and challenges were, their own words were even more captivating. In successive films, the astronauts became a more vocal presence, sharing their own personal reflections to complement the scripted narration. Personalities and perspectives gave audiences new insight into the spaceflight experience and perhaps also into commonalities that enabled viewers to think "that could be me." The shuttle era had been heralded as the age when spaceflight might become possible for ordinary people. Although actual flights of nonastronauts were few, through the IMAX films, millions of ordinary people "went" into space. And still do. Viewers now enjoy the IMAX spacefaring films in the comfort of home on a flat screen television or computer monitor via high-definition DVDs with Dolby digital sound, in a choice of languages and with a menu of interviews and other bonus features. Sales remain brisk online and in IMAX venues.

Since Michael Collins' 1976 proposal to put an IMAX camera on the shuttle through release of this six-film series, the constant goal was to present spaceflight as the astronauts experienced it and show Earth as they saw it. A corollary was to let the spacefarers be themselves, not to shape them into types or icons or idealized heroes. During the thirty years of shuttle spaceflight, commercial filmmakers never attempted a realistic treatment of this rich subject. When the shuttle rarely appeared in Hollywood films, the plots were improbable, and the astronauts were caricatures.[53] Whereas *Apollo 13* (1995) won deserved praise for recreating the reality of that mission, IMAX films captured reality directly, without recreation. In the court of public opinion as well as critical reviews from many quarters, the IMAX films accomplished their mission. To the astronauts belongs the ultimate credit for bringing spaceflight down to Earth.

Notes

IMAX® and *The* IMAX *Experience*® are registered trademarks of IMAX Corporation.

1. This film will be referred to as either *The Dream Is Alive* or *The Dream* throughout the essay. Total recorded theater attendance for the first six months after release of *The Dream Is Alive* totaled 2.6 million in 17 venues in the United States and Canada. By 2012, 42.6 million viewers had seen *The*

Dream in theaters worldwide, and total attendance for the six filmed-in-space IMAX documentaries reached almost 100.7 million; IMAX correspondence with the author, October 2012.

2. Footage shot on the space shuttle at the same time in Cinema 360, a 35mm dome-screen format, resulted in a program shown in several planetariums.

3. The primary source of records of IMAX space films is Graeme Ferguson and his wife Phyllis Ferguson. He coinvented the IMAX technology, cofounded IMAX Corporation, served as its CEO, and directed and produced many IMAX films on various subjects. The space films became his special project for more than thirty years, and his imprint is on all of them. Both Fergusons were actively involved in every technical, artistic, and administrative aspect of *The Dream Is Alive* and its successors, and their files contain almost daily accounts of the progress of the projects. This account draws heavily from their papers, certain ones of which are cited, as well as an oral history project with Graeme Ferguson and the author's series of interviews with both Fergusons in 2010–2012. The narrative arc of *The Dream* project is so pervasive that it cannot easily be attributed to particular documents. Instead, this blanket acknowledgment of the Fergusons as the primary source of *The Dream Is Alive* story must suffice. Other sources are cited separately.

4. In the first years of shuttle flights, NASA issued a series of "Going to Work in Space" posters and various public outreach materials with similar titles. The theme of working in space permeated the shuttle era from President Nixon's 1972 statement on the decision to develop a shuttle to the final shuttle mission in 2011.

5. IMAX Corporation provided four cameras for spaceflight. On the 1984 missions, only an in-cabin camera flew; then in 1985 a second camera was mounted in the payload bay for wide-angle shots of the Earth and extravehicular activity. In 2000 IMAX replaced these 2-D cameras with two 3-D cameras, one in the cabin and one in the payload bay. Astronauts operated all cameras, those in cabin in hands-on mode and the ones in the payload bay by remote control.

6. Vincent Canby, "'Big Screen' Takes on New Meaning," *New York Times*, April 19, 1987.

7. Glenn Whipp, review of *Hubble 3D*, *Los Angeles Times*, March 19, 2010.

8. Roger Ebert, "*Hubble 3D*: A Journey into Time and Space," *Chicago Sun Times*, April 21, 2010; Jennifer Merin, "Heading for Hubble and Beyond," http://documentaries.about.com/od/revie2/fr/Hubble_3D_Movie_Review.htm (accessed March 15, 2012).

9. Justin Chang, review of *Hubble 3D*, *Variety*, March 14, 2010.

10. Michael Collins, Director, Smithsonian National Air and Space Museum, to Dr. Joseph P. Allen, NASA Assistant Administrator for Legislative Affairs, November 17, 1976, acknowledged by Allen's return letter dated December 20, 1976, Smithsonian Institution Archives, Record Unit 338, Box 4, IMAX file, Washington, DC (hereafter referred to as SIA).

11. A similar IMAX theater system called Omnimax, modified for projection on a planetarium dome, debuted a few months earlier in 1973 at the Reuben H. Fleet Space Theater and Science Center in San Diego.

12. Recounted by Graeme Ferguson in IMAX Oral History interview by David Cobb and Loral Dean, November 7–8, 2000, transcript, pp. 90–91. Graeme and Phyllis Ferguson, personal collection.

13. The National Air and Space Museum built the third institutional theater, following the Reuben H. Fleet Space Theater and Science Center in San Diego (1973) and the Living History Center in Philadelphia (early 1976). Two of the six theaters were in theme parks (Circus World in Orlando, 1974, and Cedar Point in Sandusky, Ohio, 1975), and the other was the temporary theater in the Spokane World Expo (1974). Graeme Ferguson, communiqué to author, December 20, 2011.

14. Graeme Ferguson, President, IMAX Systems Corporation, to Chester M. Lee, NASA Director of STS Operations, July 5, 1978, SIA, Record Unit 338, Box 4, IMAX file.

15. Memorandum from NASA Director, STS Operations, to Chief, Launch Agreements and Customer Services, "Position on Filming Aboard the Shuttle," October 4, 1978, SIA, Record Unit 338, Box 4, IMAX file.

16. Graeme Ferguson and Toni Myers, correspondence with author, March 27, 2012.

17. "*Hail Columbia!* Featured at NASA's 25th Birthday Party," *IMAX Projections Newsletter*, January 1984, SIA, Record Unit 338, Box 13, IMAX file; Ferguson, IMAX Oral History, transcript pp. 126–29.

18. The IMAX films produced by the four-way collaboration of IMAX, the National Air and Space Museum, NASA, and Lockheed were *The Dream Is Alive* (1985), *Blue Planet* (1990), and *Destiny In Space* (1994). IMAX and NASA cooperated in producing three more: *Mission To Mir* (1997), *Space Station 3D* (2002), and *Hubble 3D* (2010), with Lockheed sponsoring *Mission To Mir* and *Space Station 3D*. (Warner Bros. Pictures cofinanced and codistributed *Hubble 3D*.) Toni Myers of IMAX wrote all of these films, edited all but one of them, narrated *Blue Planet*, and directed and produced both 3-D films. Graeme Ferguson directed *The Dream Is Alive* (and *Hail Columbia!*) and produced or coproduced all of the space films.

19. Each of the films opened with a statement screen, "Presented as a Public Service." According to Graeme Ferguson, the project team never anticipated that the films would be profitable because filming in space was far more expensive than regular IMAX shoots and there were at the time too few theaters to turn a profit. While successful films would fill those theaters and bring revenue to the museums and parks, IMAX and Lockheed hoped at best to break even and earn back the production costs. Lockheed decided that the only feasible way to fund successive films was to roll over any proceeds from one into producing the next. Over time, Lockheed did not quite recover its initial investment but continued to view sponsorship as a public service. The ultimate goal of the IMAX Corporation in the collaboration was to grow the market for new theaters, not to make a profit on the films per se. Graeme Ferguson, interview with author, January 28, 2012.

20. The title came from John Young's comment at the STS-1 postmission press conference, spoken near the end of *Hail Columbia!*: "The dream is alive again. Let's keep it that way."

21. Ferguson, IMAX Oral History, 131–32. IMAX had thought about modifying the camera for use in space, but astronaut Bruce McCandless advised that they would rather work with a tried-and-true device than run the risk of problems with a new one. This anecdote reported in the *Baltimore Evening Sun* review by Steve McKerrow was mistakenly attributed to astronaut Joe Allen rather than Bruce McCandless (June 20, 1985).

22. STS 41-C in April, STS 41-D in August–September, and STS 41-G in October, all in 1984. These were the eleventh, twelfth, and thirteenth shuttle missions, with *Challenger* and *Discovery* flying five times that year.

23. "A General Statement of Conditions and Proposed Letter of Agreement," undated and unsigned on National Air and Space Museum letterhead; probably composed by Director Walter Boyne or the museum's Public Affairs Chief Brian Duff in 1983. SIA, Record Unit 338, Box 23, *Dream Is Alive* file; *The Dream Is Alive* agreement with Lockheed and attached narrative, July 31, 1984, SIA, Accession 04-092, NASM, Exhibits Production Records, *Dream Is Alive* file.

24. "A General Statement of Conditions and Proposed Letter of Agreement," SIA, Record Unit 338, Box 23, *Dream Is Alive* file.

25. The story of these preparations is recorded in an IMAX presentation "Preparing the IMAX Camera for Space." Graeme and Phyllis Ferguson, personal collection.

26. Museum Director Walter Boyne to IMAX President Graeme Ferguson, June 27, 1984; Museum Associate Director for External Affairs Brian Duff to NASA JSC Director of Public Affairs Harold Stall, July 13, 1984; Brian Duff to *The Dream* team members, memorandum, July 20, 1984; Lockheed Vice President H. David Crowther to NASA Administrator James M. Beggs, July 2, 1984. All in SIA, Record Unit 338, Box 23, *Dream Is Alive* file. Also see letters of appreciation from Museum Director Walter Boyne to the crews of shuttle missions 41-C, 41-D, and 41-G, October 2, 1984, SIA, Record Unit 338, Box 13, *Dream Is Alive* file.

27. Evidently, there was never an effort to fly an IMAX expert on the shuttle as a payload specialist to do the filming. The IMAX training team was convinced that the astronauts were highly capable for the task, and this confidence was amply rewarded by spectacular camera work. Ferguson, IMAX Oral History, 152, and correspondence with author, December 9, 2011.

28. Toni Myers, IMAX writer/editor/producer, telephone interview with author, November 18, 2011; "Adventures in Space," special feature produced in association with Creative Domain, *Space Station*, DVD, directed by Toni Myers (Burbank, CA: Warner Home Video, 2005). IMAX cinematographers, notably James Neihouse and David Douglas, also led astronaut training sessions and participated in the production of the films, as did sound experts Ben Burtt and Greg Smith. Burtt also served as principal director of *Blue Planet*.

29. Myers, telephone interview with author; "Adventures in Space."

30. Toni Myers and Marsha Ivins, "Commentary," *Space Station*; Ferguson, interview with author. Film rolls shot three minutes of footage for the in-cabin camera and eight minutes for the cargo bay camera; Graeme Ferguson, correspondence with author, March 31, 2012.

31. Ferguson, interview with author.

32. Marsha Ivins, "Commentary," *Space Station*.

33. *The Dream Is Alive* storyboards, IMAX Systems Corporation, 1983, Graeme and Phyllis Ferguson, personal collection.

34. "*The Dream Is Alive*—Narrative," attached to film project agreement, SIA Accession 04-092, *The Dream Is Alive* Agreement.

35. Correspondence between Museum Director Walter Boyne and CBS News chief Walter Cronkite, May–June 1983, and between the museum's Associate Director for External Affairs, Brian Duff, and Mr. Cronkite's agent, Tom Stix, August–October 1984, and from IMAX President Graeme Ferguson and Walter Boyne in February and August 1985, SIA, Record Unit 338, Box 23, Cronkite file.

36. There were five astronauts per mission, but Robert Crippen flew twice as commander (STS 41-C and STS 41-G). In addition, three payload specialists who had no IMAX duties were aboard: Charles Walker (STS 41-D) and Marc Garneau and Paul Scully-Power (STS 41-G).

37. Charles W. Smith, "It's Colossal! It's Stupendous! It's IMAX!" *Reader's Digest*, August 1985, 87; Ferguson, IMAX Oral History, transcript, p. 152.

38. "Inside IMAX *Hubble 3D*," *Hubble*, DVD, directed by Toni Myers (Burbank, CA: Warner Bros. Entertainment Inc., 2010).

39. "Adventures in Space."

40. See the chapter by Matthew H. Hersch in this volume.

41. Joseph D. Atkinson Jr. and Jay M. Shafritz, *The Real Stuff: A History of NASA's Astronaut Recruitment Program* (New York: Praeger, 1985).

42. Amy E. Foster, *Integrating Women into the Astronaut Corps* (Baltimore: Johns Hopkins University Press, 2011).

43. See the chapter by Jennifer Ross-Nazzal in this volume.

44. The voices of Charlie Bolden and Jim Buchli speaking extemporaneously about their impressions of Earth from space were identified in the narration.

45. Footage for *Hubble 3D* was shot by the crews of STS-31 (1990), STS-61 (1993), and STS-125 (2009). The STS-125 crew shot 3-D footage; footage from the two earlier flights, shot for *Destiny In Space*, was converted to 3-D for use in the Hubble film.

46. Toni Myers' comment in notes from the October 2, 1985, planning session. Graeme and Phyllis Ferguson, personal collection. She always advocated for the human connection between audience and astronauts' experience; Toni Myers, interview with author, November 11, 2011.

47. Brian Duff, NASA Associate Director for External Affairs, to Stephen Chaudet, Lockheed Corp. Director of Public Affairs, May 6, 1986, SIA, Record Unit 338, Box 22, *Dream Is Alive* file.

48. This is an arguable assertion in light of the high-risk Apollo lunar missions.

49. Footage for *Space Station 3D* was shot by the crews of STS-88 (1998), STS-92 and STS-97 (2000), STS-98, STS-102, STS-100, and STS-104 (four successive missions in 2001), and ISS Expeditions 1 and 2 (2000–2001).

50. Several of the *Space Station 3D* participants talked about these matters in "Adventures in Space."

51. *Space Station 3D* held second place with 20.1 million viewers in its first ten years. These figures do not include VHS and DVD sales. IMAX correspondence with author, October 2012.

52. *The Dream Is Alive Three-Year Report* (IMAX Corporation Document, 1988) includes a digest of reviews. Praise from NASA Administrator James Beggs and others eased the way for the following films.

53. Notably, *Moonraker* (1979), *SpaceCamp* (1986), *Armageddon* (1998), *Deep Impact* (1998), and *Space Cowboys* (2000). See the chapter by Matthew H. Hersch in this volume.

CHAPTER 8

You've Come a Long Way, Maybe:
The First Six Women Astronauts and the Media

Jennifer Ross-Nazzal

On January 16, 1978, National Aeronautics and Space Administration (NASA) Administrator Robert A. Frosch released the names of the first class of space shuttle astronauts. They included—for the first time—six women and four minorities (Figure 1). Nicknamed the Thirty-Five New Guys, the astronaut candidates often joked that their class included "ten interesting people and twenty-five standard white guys."[1] The media attention the women received was staggering, as editors and reporters from popular magazines and newspapers jumped on the story, as did TV news outlets from across the globe.[2] Sally K. Ride, then a PhD candidate at Stanford University, participated in her first press conference that day. Writers and photographers from *People Weekly* soon appeared at offices and on the front steps of the astronauts' homes, and telephones rang off the hook with inquiries from the major television networks. The inclusion of women in the astronaut corps prompted some to question how the agency might have evaluated the candidates.

Bill Hines, a reporter from the *Chicago Sun Times*, asked Johnson Space Center (JSC) Director Christopher C. Kraft and Frosch how Shannon W. Lucid, the oldest woman and mother of three, would balance her career and home life. Had NASA considered her responsibilities as a mother? Recognizing that feminists might be put off by his insinuation, Hines prefaced his thoughts by admitting it was a "male chauvinist pig" question; he made no mention of how the newly selected male astronauts, many former test pilots, might balance their family duties and sixty- to eighty-hour work weeks.[3]

Although women had made gains by 1978 with the second wave of the women's movement and the passage of civil rights legislation, being an astronaut had been a male domain, so the media provided a reassuring message that emphasized the feminine traits of the first six women astronauts. Publications about their selection, training, and first missions narrowed in on the topics of

FIGURE 1. NASA's first six women astronauts pose for a picture with an Apollo Extravehicular Mobility Unit. From left to right: Shannon Lucid, Rhea Seddon, Kathy Sullivan, Judy Resnik, Anna Fisher, and Sally Ride. NASA Image S79-29596, courtesy of NASA.

weight, appearance, diet, exercise, domesticity, dating, marriage, and motherhood. On the basis of unspoken gender assumptions, these essays reinforced the belief that men and women astronauts were different, leading some to think that the inclusion of women in the corps was simply a publicity stunt on the part of NASA. It did not help that sometimes the media, NASA Public Affairs, and other astronauts—especially their mission commanders—called them "girls" and "ladies." The media even scrutinized their looks. Called the "glamournauts" and "space queens," the women could not escape discussions about their appearance or beauty they supposedly brought into the office and to their crews.[4]

In this chapter I argue that issues about their femininity, motherhood, domesticity, emotions, weight, and appearance continually dogged these six because of their sex. Competing images of the women emphasized their toughness and exceptionalism and at the same time reaffirmed their femininity.[5] In the mid-1980s the inclusion of payload specialists from Congress and other nations as well as the loss of *Challenger* drew the media's attention away from the women.

My research also demonstrates that coverage of each female astronaut candidate varied, depending upon how well she reflected traditional views of womanhood and her attitudes toward the media. The more feminine and open a woman was to the media corresponded directly to the amount of attention she received until NASA named America's first woman in space. The photogenic Anna L. Fisher received an overwhelming amount of interest initially because she met American society's standard of beauty, and she was happy to sit for interviews; in 1978 she was newly married and later became the first mother to fly in space. She appeared on multiple magazine covers even after NASA selected Ride as the first American female to fly in 1982. Ride eventually replaced Fisher as the Astronaut Office's cover girl, although the press found her to be much more prickly than Fisher. Previously, Ride had been seen as a tomboy, with a passion for sports, and therefore of less interest to the media. Lucid and M. Rhea Seddon were not only astronauts but mothers and wives, and the media frequently documented their ability to combine work and family. Publicity for the two women who remained single, Judith A. Resnik and Kathryn D. Sullivan, was sparse until Resnik became the second American woman to fly and Sullivan became the first American woman to do a spacewalk. Resnik was a very private person, which contributed to her lack of coverage, whereas Sullivan was considered, out of all the women, too boyish.

Gender, the Media, Women, and Feminism

It is a truism that women are evaluated first on their dress, appearance, and overall femininity and then finally by their actions or statements. Over the past forty years numerous social scientists have reached this conclusion. Newspaper reports, magazine articles, and television coverage of women generally focus more on their feminine attributes than their accomplishments or ideas, especially when compared to their male colleagues.[6] The media tends to rely heavily on stereotypes to the detriment of their female subjects, who are viewed by audiences or readers as novelties, not real contenders in their fields.[7]

Studies demonstrate again and again that the coverage of men and women in magazines and newspapers in the late 1970s and early 1980s changed little even as the women's movement raised public awareness of key issues facing women. Marjorie Ferguson's book *Forever Feminine: Women's Magazines and the Cult of Femininity* is one such example. Her study found that the content of British women's magazines in the late 1970s did differ on the surface—as women increasingly worked outside of the home and the stigma associated with leaving children at day care diminished—but the major theme of women's weeklies remained generally unchanged. Women's magazines, she concluded, promoted a "cult of femininity," which encouraged women to believe being a wife and mother was the most important goal of their lives.[8]

More recent studies of these types of publications concur with Ferguson's findings. Tough women and girls featured in women's magazines, Sherrie Inness argues, are "only skin-deep." Women's magazines show only women with masculine traits "that can be toned down" to assure "readers that displays of toughness pose no serious challenge to the cult of femininity that the magazines uphold."[9]

Newspapers demonstrated similar biases in the 1970s and early 1980s, which suggested that the feminist movement had made little, if any, headway in changing perceptions about women's role in sports, government, or business. Junetta Davis's 1979 study of eight newspapers, four from Oklahoma and two each from the East and West Coasts, found large disparities between coverage of men and women in these dailies. In the sports, editorial, or business sections, men were the focus. Women appeared mainly in their own dedicated section, which editors devoted to cooking, home decorating, family, and women's organizations. There were other significant differences. Reporters tended to physically describe women's attributes and identify them as wives or mothers. Some of Davis's generalizations can be applied to the first six women astronauts. Newspapers, she determined, overlooked American women, with only 8.6% of the articles including a woman as a main character. Given the feminist movement and the important role of the female consumer, she asked, "Can newspapers ignore reality" and fail to keep pace with women's shifting social and economic roles in the United States?[10]

Roy E. Blackwood was similarly critical of the *Washington Post* and *Los Angeles Times* for featuring male photographs more frequently in the sports, business, and news pages from July 1980 to June 1981. Photographs of professional women, which appeared mainly in the lifestyle or obituary pages, did not reflect the advances made by women in higher education, government, or industry. "The result," Blackwood wrote, "is that their photo coverage is more removed from reality than it was nearly a decade ago," before the women's movement reached its apex.[11] W. James Potter's study of elite American newspapers like the *New York Times* and the *Chicago Tribune* found similar trends documented by Davis just a few years earlier. Journalists rarely included women in articles about business, government, or science. Instead, they were more likely to be featured in people-oriented articles, crime features, or accident coverage.[12]

The trends documented by researchers in the 1980s suggest that change, in spite of the women's movement, moved at a glacial pace. Although legislative and legal victories for women could be documented, American women continued to encounter barriers. Some attitudes had changed, but not everyone's opinions about women, career, and family had evolved, as evidenced by two polls taken near the end of the 1970s. Feminists had argued against the old adage that a woman's place was in the home, but in 1977, just one year before NASA selected the first class of shuttle astronauts, polls showed that most Americans overwhelmingly agreed with this statement.[13] A 1978 study indicated that

Americans were less likely to vote for a female candidate even though Americans had previously elected women to local, state, and federal positions.[14]

Traditionalists worried about the impact the women's movement might have on family life and marriage, and an antifeminist backlash surfaced in the 1970s as the right-to-life movement grew and opponents joined the campaign to stop the Equal Rights Amendment.[15] There were other significant hurdles. As American women broke free of legal constraints and gained greater reproductive rights, the social control they previously encountered shifted to beauty standards. Women were expected to appear thin, beautiful, and youthful. Naomi Wolf argued that these images, which she called the "beauty myth," were used "as a political weapon against women's advancement."[16]

Although women had finally made it into the Astronaut Office in 1978, women astronauts were still grounded, as demonstrated by the media's reliance on female stereotypes.

Weight and Appearance

From the beginning the media had no qualms about discussing the weight and appearance of the newest members of the astronaut corps, perhaps because they were women or because they were celebrities. Women's, news, and entertainment magazines, as well as newspapers, regularly noted the weight and height of the first six women because a woman's measurements equated with one's femininity and beauty.[17]

For example, when the 1978 class visited the National Space Technology Laboratories (now called the Stennis Space Center) in Mississippi, the local newspaper talked with four members of the thirty-five astronauts: Ride, Fisher, Guion S. Bluford, and David M. Walker. A photograph of Fisher, labeled "pretty astronaut," appeared in the local paper. The caption described the brunette with "hazel eyes"; she "stands five feet, four inches tall and weighs 110 lbs," the prime example of womanhood.[18]

In May 1979, *Weight Watchers* featured a glamorous Hollywood-like shot of Seddon in her blue flight suit, hand draped over her helmet, on their cover. The first few lines of the article focused on her appearance, "She looks like a college cheerleader—one of the nicest and prettiest girls on sorority row. She's petite, blonde, and seems too young to have had such a past; too soft to have such a future."[19]

Whereas Fisher and Seddon received the bulk of the media's attention for typifying femininity and for their ability to maintain their slender frames, Sullivan received less attention presumably for the opposite reason. As she concluded, she was too much of a "tomboy, too smart, too strong, too all those things"; she was "not the archetypal little girl" compared to the other women the media could feature on their covers.[20] Sullivan was taller and heavier than the other women in the office, something newspapers and magazines regularly mentioned. A 1984 article from the *Washington Post* pointed out that Sullivan, then 150 pounds, was "the most robust of the women astronauts" in the office at the time.[21] Plus, she played rigorous sports like ice hockey. Whereas Fisher and Seddon looked like Olympic gymnasts in magazine photos, Sullivan was much more rugged in appearance.

Diet, Exercise, and Fitness

Fisher and Seddon's diet and exercise routines received more coverage than those of the other four women selected in 1978, but there was an overall interest in all of the women's general fitness. Coverage of the women's workouts and diets varied, depending upon the type of publication. Celebrity magazines like *People Weekly* mentioned their exercise routines prior to their selection and arrival in Houston.[22] The articles contained within *Redbook: The Magazine for Young Women* and *Weight Watchers* highlighted Fisher and Seddon's daily workouts and slender frames and provided their fitness and diet tips. These essays fit the themes of women's magazines, which generally emphasize "how to become healthier, fitter, thinner, and more attractive," with an emphasis on "looking good."[23] And dieting exemplifies a woman's femininity.[24] Although dieting and exercise are often promoted as part of a healthy lifestyle, the purpose is to meet societal and cultural standards of beauty that are frequently criticized as unobtainable.

Articles about the female astronauts' exercise routines and diet continued to follow the women for much of the early years of the shuttle program, and this fascination related to America's interest in achieving the perfect physique in the 1970s. Plus, NASA expected its astronauts to keep in shape, so magazines hoped to learn the secrets of how these women, particularly Fisher and Seddon, kept slim.

In June 1978, Fisher appeared on the cover of *Redbook*. She was still working as an emergency room physician in California, so *Redbook* discussed her fitness tips—and NASA's suggestions for body conditioning—in that issue. Fisher's "plan for staying in shape can work for you—whatever your orbit," the magazine promised. Although the article noted Fisher's achievements as one of the first women astronauts, it also touted the benefits of exercise and diet for the doctor: "good looks, high energy, and the stamina to cope with stress."[25] The emphasis was first and foremost upon physical attractiveness over other potential health benefits associated with exercise like lower cholesterol rates. A photograph of a slender Dr. Fisher in a track suit graced the first page of the article. Weighing 105 pounds, Fisher represented the traditional ideal of American womanhood; readers, *Redbook* suggested, could achieve the same results by becoming more active. Writer Andrea Fooner explained that Fisher's approach was easy for any working woman to attain.

More than thirty years later Fisher remembered the interview. The magazine wanted to tell readers how she stayed slim and trim and kept pressing her on the issue. "I'm trying to say politely," she recalled, "what I couldn't. 'Hey, look, I just got off of doing twenty-four-hour shifts for the last couple of years. My workout regime is eat dinner and sleep.'"[26] She really did not have any tips; she was busy and being run ragged. Fisher did, however, explain how she began training to become an astronaut. She and her husband, who was also at the interview and photo shoot, ran several miles a day to prepare for NASA's physical exam. Exercise tip number one related to running. The Fishers, both physicians, endorsed running as "the most efficient way to improve cardiovascular functioning and over-all body conditioning."[27] To emphasize that point, the magazine featured Fisher in a jogging suit, running in place at Ron Colby's New York photography studio.

Another tip related to food consumption and diet. How could a "working girl" maintain her weight when her refrigerator typically only contained ice cream or a frozen cheesecake and she ate out every day? The Fishers, readers learned, ate lean meats like poultry and fish and consumed yogurt, fresh fruits, and vegetables. Instead of consuming a greasy hamburger for dinner, the couple preferred a meal of Japanese cuisine because of the protein, vitamins, and minerals. The instructions were simple for any reader hoping to look like twenty-nine-year-old Fisher: eat sensibly—choose healthier meats, fruits, and vegetables—and run two miles a day. If an emergency room doctor could do it, *Redbook* reasoned, an average woman working eight hours a day juggling family, career, and home certainly could too! As Fooner concluded, any woman could emulate Fisher's "sense of active fun" to achieve her weight loss and diet goals.[28]

A similar article about Seddon appeared in *Weight Watchers* about a year later. Called "The Lady Astronaut's Diet," the article explained how Seddon maintained her weight as an astronaut, providing examples of her exercise routine—running two miles a day—the types of food she ate, and how she topped her vegetables and salads. If readers had missed Fisher's tips in *Redbook*, they could simply follow the same basic formula that worked for Seddon. If Seddon could lose ten pounds while working as a medical intern following the Weight Watchers program, so could readers. This message, coming from "an expert in the field of nutrition" and a physician who knew about the benefits of diet and exercise, did, however, override the emphasis on beauty found in other articles about diet and exercise.[29]

In 1980 *Redbook* asked Fisher to appear in another physical fitness essay, called "Exercise with an Astronaut." The magazine took Fisher's typical workout routine and met with Dr. Michael A. Berry at JSC to come up with a plan for *Redbook* readers who wanted to strengthen their upper body. The magazine's Beauty Department, not the Fitness Department, put together this special piece.[30]

Wearing a specially designed red leotard and tights with blue and white stripes down the side, Fisher was the epitome of the feminine form at the age of thirty. She was not flabby but slender with no bulging muscles. At the time bodybuilding was associated with men, not women, and readers were specifically told they would not develop any bulk with this workout. Women's hormones prohibited the development of well-defined muscles, and the photos of Fisher provided the reassurance women needed.[31] Berry had delivered on his promise that the female astronauts would not become men or look manly in appearance by becoming astronauts and exercising with weights to build upper body strength. In 1978 he jokingly told *House and Garden*, "We wouldn't want them to look like Arnold Whats-his-name," and Fisher did not.[32]

Domesticity

Given the belief that men and women were different, the press expressed an interest in the women's homemaking skills. Relying on stereotypes that a woman's most important job in life was to be a wife, mother, and homemaker, the media asked the women about their culinary skills and interest in housekeeping. The six, reporters assumed, found housekeeping rewarding, but journalists overemphasized the presumed link between gender and domesticity. The press tended to overlook the fact that these women were not just mothers and wives. They were scientists, engineers, physicians, and astronauts. Nonetheless, at one point they classified half of the women as gourmet cooks but did note that they had little time to devote to the task. Publications regularly mentioned and passed judgment on the women's ability to keep a house, office, or apartment clean.

In 1978, just after their selection, *People* reported that several of the women were accomplished in the kitchen. Resnik was a gourmet cook, and Seddon loved to bake "lavish desserts." Ride, by contrast, appeared less skilled in domesticity. *People* insinuated she was a poor housekeeper with a "cluttered campus office and littered apartment."[33]

One might expect such categorization from entertainment magazines but not from hard-hitting news magazines or newspapers. Yet even these publications highlighted women's domestic skills. *U.S. News & World Report*'s Paul Recer described Lucid as a typical suburban mom in 1981—"a 38-year-old matron with three children, a house in the suburbs and a fondness for gourmet cooking"—with one small difference, she was an astronaut. But, according to one member of the Astronaut Office, she was not "your typical mother of three. She can talk numbers with the best engineers anywhere."[34]

Recer's categorization of Lucid, a gourmet cook, differed dramatically from a 1978 article in the *Dallas Morning News* that suggested the biochemist was too busy to cook three-course meals. After all, she had three young children and worked full time. Cooking of that type required extensive planning and time in the kitchen. When asked about her culinary talents, Lucid jokingly replied, "Well, we're all alive." Her best dish? "Whatever's quick, easy and leaves no mess." What was her husband's favorite meal? Her mother's roast, she kidded. Lucid, the *News* noted, did not enjoy cleaning the bathrooms or kitchen of her Oklahoma home. "But she is typically feminine," the author noted. Upon seeing the bland beige interior of the orbiter she thought of ways to add color to "brighten up the place."[35]

Dating and Marriage

A woman's femininity has often been defined by whether or not she is married. Although feminists did not think a woman necessarily needed a man to be happy nor did she need to be defined by her ties to men, the media in the late 1970s, particularly women's magazines, continued to perpetuate the idea that a woman needed a man to complete her.[36] Two of the women were married when they were selected, and the media showered both with attention. By contrast, the others were regularly asked about their status. Lucid had been married for more than a decade, whereas Anna and William F. "Bill" Fisher were newlyweds. The selection of two married women reflected a role reversal of sorts, as women generally followed their husbands, the family breadwinners, when they changed jobs, not the other way around.

The Lucids met when Shannon Wells applied for a job at the Kerr-McGee Corporation, where Michael Lucid worked; he thought she was overqualified for the position, but his boss disagreed and hired her. They married a year later. Shannon had always been interested in becoming an astronaut, and Michael supported her dream by telling her that he would quit his job and move with her. "It sounded like a perfectly valid goal to me," he told *Oklahoma Monthly*. The idea of his wife becoming an astronaut did not seem strange or impossible. "As a scientist," he explained, "I deal with the far-fetched every day." Asked about moving south to Texas to follow his wife's career, Michael said, "It's a once in a lifetime opportunity. I don't see anything wrong with it."[37] Most men would have insisted that their wife decline the position rather than move so that their wife could have her dream job, and a majority of American women agreed.[38]

Bill Fisher was similarly supportive of his wife. He had also been a finalist for the position of mission specialist that year but did not make the final cut, although NASA chose him two years later. Many of the early articles about Anna focused on her selection and the fact that NASA had not included Bill on the list of thirty-five candidates. Nonetheless he, like Michael Lucid, saw this as the opportunity of a lifetime for his wife. Not everyone believed Bill's assertion that he was not jealous of Anna. A local California TV talk show, featuring both Anna and Bill, focused on the 1978 decision. The male host tried to get Bill to admit he may have been disappointed with the outcome. "When you're alone in your car driving to work, don't you ever pound the steering wheel?" Upon hearing no, the interviewer asked, "Well, just between you and me, do you squeeze it a little?"[39]

With only two married women astronauts selected, the press highlighted the dating habits of the other four. Their interest in romantic relationships proved that they were "normal" women despite the fact that they were astronauts. Ride had a serious boyfriend who would move to Houston with her, whereas Seddon, who worked eighty hours a week as a surgical resident, casually dated. By contrast, Resnik lived alone in a Redondo Beach singles complex. When asked if she had any concerns that being an astronaut might negatively impact her personal life, Ride responded, "The space program is going to have absolutely no effect on my plans for marriage." Seddon admitted to being "single by accident. My career has become more exciting than most men."[40] Both later married men from the 1978 class, but interest in these astronaut couples, when compared to the Lucids and Fishers, was relatively light.

Motherhood

Although opportunities for women had expanded outside of the home in the 1970s, women's weeklies continued to emphasize the importance of raising children.[41] "New momism," which came into vogue in the 1980s, fostered the belief that no woman could be happy unless she had children, raised them herself, and physically and emotionally devoted herself to her kids. The idea played to the stereotype that women "are, by genetic composition, nurturing and maternal, love all children, and prefer motherhood to anything, especially work." Intensive mothering became romanticized just as more women were working outside of the home and were unable to dedicate the time and effort to meet these impossible ideals.[42]

Astronaut moms challenged this popular ideology. They refused to sacrifice their careers to stay at home and raise their children. Their decision to work challenged traditional family values, and some people questioned their choices. But the three mothers wanted to be astronauts and combine work and parenting. Motherhood was a part of their lives, but raising kids was not the sole source of their happiness; their careers also fulfilled them.

Lucid, as noted earlier, had three children when NASA selected her as an astronaut, and the media regularly mentioned this. Every article printed about

Lucid in 1978 identified her as the mother of three, even *Chemical and Engineering News* from the American Chemical Society.[43] Dubbed the first mother in space well before anyone in her class had flown, the media was curious about her, and they traveled to Oklahoma City in 1978 to document her life. Portraits of the Lucids depicted a traditional nuclear family with three happy children and a doting mother and father.[44]

Her selection brought up lingering questions that working mothers with children still face. Was being an astronaut not a dangerous job? Some of America's best and brightest test pilots who became astronauts died in T-38 crashes in the 1960s, and the crew of Apollo 1 lost their lives in a fire at the launch pad at the Kennedy Space Center (KSC). Didn't members of the office work long hours and spend days away from home? Countering potential objections that being an astronaut was simply too risky for a mother or too time intensive, Lucid admitted that she had no fears about how her family would adapt because she had "a very capable husband" who could raise the children.[45] But the question would not go away. Even two years after Lucid was named to the office, Judy Stein, a writer for *Texas Woman*, wondered how the job had affected Lucid's family. Being an astronaut had no detrimental impact on her kids because, as she noted, "I've always worked, so this job hasn't really affected the children."[46] The motherhood topic faded for Lucid as other women in the class had children, but questions about "bad mothers" continued.

In the spring of 1982, Seddon and her astronaut husband Robert L. "Hoot" Gibson, who she married in 1981, issued a short press release announcing that they would be expecting a baby later that summer.[47] The press seemed to be perplexed with how to handle the one-line announcement. They had never dealt with a pregnant astronaut before, just expectant astronaut fathers. Naturally, they expected that Seddon, who had not changed her name following her wedding, would take Gibson's last name as is convention in most American families, and they referred to her as Rhea Seddon Gibson. She did not, however. Coverage of the announcement focused primarily upon the first pregnant astronaut, whose official NASA portrait appeared in many of the news briefs, not the father.[48] Interest in Seddon's pregnancy continued that summer as photos of the expectant mother, dressed in her maternity clothes, appeared in newspapers across the country.

The birth of Paul Gibson sparked as much interest as their announcement. Their son, known as the first astrotot, suffered some breathing difficulties after birth and was transported to Houston's Medical Center. After his release, the couple participated in a press conference to talk about Houston's Life Flight helicopter and Hermann Hospital's neonatal intensive care unit. The reporters showed great interest and asked if Seddon would return to work. And, now that she was a mother, would she still be eligible for flight? "NASA," Seddon later recalled, "was happy to show life goes on. It's normal for our women astronauts, and we're all for [motherhood]."[49]

Even the arrival of NASA's second astrotot was big news in the summer of 1983. Again, much like Seddon, Fisher—not her husband Bill—received the bulk of the attention from the media, with her portrait also included in the press briefs. Much was made of the fact that Fisher returned to work on Monday after her Friday delivery.[50] She remembered the events as they unfolded: "Monday, I was just feeling so happy and so good. I was assigned to a flight. I had my new baby. I decided to go into the Monday morning [astronaut] meeting . . . Just to say, 'I'm here and nothing's going to change.'"[51]

Once these two had children, the media frequently celebrated these mothers, and images of Seddon and Fisher with their children abounded. The new astronaut mothers, just like the Mercury 7 before them, became celebrities. Paul Gibson, only five months old, appeared with his mother in a *People* magazine article about births in 1982, which featured the children of actresses Jane Seymour, Sissy Spacek, and Lindsay Wagner.[52] Only a few months after STS-7, the first mission to include an American woman, *Good Housekeeping* featured a two-page photograph of the new mothers of the office with their husbands and babies. Fisher, Bill, Seddon, Hoot, and Paul Gibson all wore the blue NASA flight suits; Kristin Fisher wore a onesie with a NASA emblem on the front.[53] In the fall of 1983, NASA announced the selection of two crews for upcoming flights. Flying American women in space was still newsworthy, so a few newspapers, like *USA Today*, highlighted the fact that NASA had just assigned two moms (Fisher and Seddon) to spaceflights.[54]

A few weeks before Fisher's first and only spaceflight, *Parade Magazine* introduced America to the soon-to-be first mother in space and posed the question "is she a good mother?" (Figure 2). Although it was becoming more acceptable for mothers with young children to work outside of the home in the 1980s, the media tended to fixate on the potential dangers of day care and other childhood perils for kids of working mothers and the nation. Some homemakers, who chose to remain at home, vilified women who juggled both home and family.[55] Bill was not sympathetic to such traditional ideas. "This is not 1900," he said. Mothers were no longer expected to stay at home until their children were grown. The finger pointing and "how dare you [leave your daughter behind]" applied equally to both parents, not just mothers. As he pointed out, mothers and fathers raised children. "No one says, 'Omigod, how could you leave your family?' to a male astronaut." Anna believed Kristin benefitted from her decision to do more than keep house and dote on her daughter.[56]

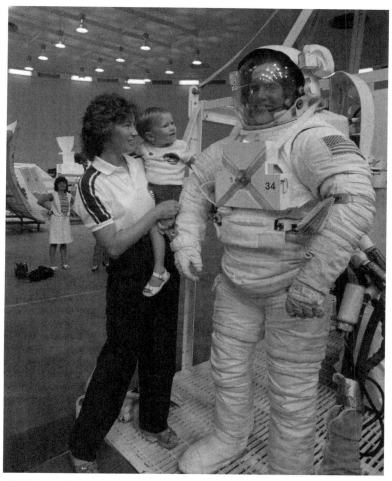

FIGURE 2. Anna Fisher and daughter Kristin look on as Bill Fisher prepares for a suit run in the Weightless Environment Training Facility. NASA Image S85-39411, courtesy of NASA.

Ladies of Space

The media also labeled these remarkable women as ladies, although in truth none of them fit this mold. They did not look like the crew-cut astronauts of earlier generations, but the women were just as remarkable as the macho astronauts of the early space programs. Even though they were only civilians, they had the grit and spirit of Mercury astronauts Alan B. Shepard and John H. Glenn. "Ladies" were prim and proper, dainty, and conciliatory. They were soft-spoken, delicate, and sometimes frail. These were not delicate women. They were "tough women": adventurous, aggressive, independent, competitive, and shrewd. The term offended Resnik, recalled John M. Fabian, who once introduced her as one of the lady astronauts. "Hell, I'm no lady!" she exclaimed.[57] That is not to say that the first six women astronauts were the exact opposite of this ideal, but to reach the heights that they did in the 1970s was no small feat. Being an astronaut was an untraditional career for a woman.

They participated in various programs that were physically demanding, like water survival training at Homestead Air Force Base in Florida. They also learned how to scuba dive, trained on ejection procedures for the T-38 jets, and flew in the backseat of high-speed jet aircraft. Photographs featured the women successfully completing these activities that Americans had once viewed as men's work. These images illustrated that women astronauts were just as powerful and capable as the men from their class, who also participated in this training (Figure 3).

FIGURE 3. Rhea Seddon demonstrates her strength by lifting classmate Robert Gibson over her head in the KC-135. NASA Image S79-33021, courtesy of NASA.

Their attitude also demonstrated their toughness. They did not believe their jobs, which entailed flying in high-speed jet aircraft and riding on rockets, were dangerous. "They're every bit as courageous and determined as the male astronauts," John Lawrence of NASA said. "That's why they were chosen."[58]

Tough women have been stereotyped as unfeminine, but this label did not stick to these six because the media simultaneously described their masculinity and highlighted their femininity; after all, astronauts, including women, had to be tough and express an interest in activities normally not associated with women. But they were still all women. Studies show that women's masculinity (an interest in masculine activities or male attributes) can be accepted and celebrated—if they fit the traditional definitions of femininity.[59] The female astronauts, who were courageous and determined, were still "true women" as illustrated by their appearance and support of traditionally feminine roles as wives and mothers.

Lucid's strength was exhibited through several anecdotes (Figure 4). Only six weeks old when she and her parents were interned by the Japanese in Shanghai, China, during World War II, she managed to emerge healthy from the camp in spite of food shortages. When the family flew in an unpressurized cargo plane to China, one of her siblings vomited, another passed out, but Lucid, intrigued by how the pilot could land on a small runway, decided she wanted to learn how to fly.[60] She did, and she eventually flew a solo roundtrip flight from Oklahoma to Alaska.[61] More than once she demonstrated her cool when the engines on her plane failed in flight, and she safely landed. Not even snakes frightened Lucid, who tied one in a knot when a male colleague tried to startle her.[62] Her family and husband were present in these articles, thereby suggesting that Lucid's essential nature reflected traditional gender mores.

FIGURE 4. Shannon Lucid descends from the top of the crew compartment trainer during emergency egress training for STS-51G. NASA Image S85-28133, courtesy of NASA.

The media illustrated Fisher's toughness through her maternity. Four months pregnant with her first child, she participated in a test to prove a woman could be an Astronaut

Support Person (ASP) at KSC. The ASP strapped in the crews for flight and also carried them out of the orbiter if there was an emergency on the launch pad. She proved women—even pregnant women—could do the job when she carried out two men playing the role of pilot and commander.[63] Despite her abilities, she was still entirely feminine; she was a wife and mother.

Even articles about Sullivan emphasized both her ladylike demeanor and bravery. A 1979 article described her as "soft-spoken" lady astronaut with little time for a personal relationship. The author also noted Sullivan's recent record setting flight in the WB-57 aircraft. In 1979, she became the first American woman to be fitted in a pressure suit designed for the high-altitude plane at Edwards Air Force Base in California. The plane regularly flew at around 60,000 feet, and soon after becoming flight qualified, she set a woman's altitude record at 63,300 feet.[64]

Missions

When NASA finally named a woman to a crew, it was in the spring of 1982. *Columbia* had flown three missions, with STS-3 successfully landing at White Sands in New Mexico. Ride would fly on STS-7, whereas her classmate, Bluford, would become the first African-American in space onboard the eighth flight of the program.[65] The announcement sparked great interest in Ride, who had been selected from the six to fly first, but reporters and writers complained she was inaccessible. She refused to sit for interviews and ignored private requests.

Later in April the four-member crews of STS-7 and STS-8 held press conferences at JSC's Teague Auditorium. (Physicians Norman E. Thagard and William E. Thornton would later be added to the crews to study space sickness.) The men wore dark blue suits, whereas the only woman on the crew opted for a blouse and skirt. Ride "looked slight and girlish" with "soft wavy hair, slightly freckled skin bearing little, if any makeup." Crew commander Robert L. Crippen, who had flown as the pilot of STS-1, introduced the only female member of the four as "the prettiest member of the flight crew" and made reference to the other girls and ladies in the office (Figure 5).[66]

Now that a woman had been selected for a flight, reporters wondered what drastic changes were in store for the space program, the crew, and vehicles. For instance, how would the "interpersonal chemistry" change with Ride as a member of the crew? How would gentlemen with "fairly traditional upbringings" deal with a woman onboard? Would men defer to women in orbit to be polite, or would these traditional gender standards, popular on Earth, disappear in orbit? In other words, would bringing women in orbit somehow make the space program less rowdy, less boisterous? Would women civilize the American space program just as they had the Wild West?[67] Following the conference,

FIGURE 5. Sally Ride and the crew of STS-7 face the media after NASA announced she would become the first American woman to fly in space. NASA Image S82-30689, courtesy of NASA.

Fabian paired with Ride as they participated in fifteen minute interviews with reporters. At the end of these sessions few felt they could answer these basic questions, "Who is Sally Ride, America's first woman in space, and what sort of person is she?"[68] Ride was more concerned with training for her flight than keeping the media happy. "I've spent my whole life not talking to people, and I don't see any reason why I should start now," she stated.[69]

About one month prior to the launch of STS-7, in spring 1983, the crew sat down with the media for their second preflight press conference. Throughout the conference reporters asked the four men much more technical questions about flying, the payloads, the orbiter, and experiments than mission specialist Ride, who tended to be asked more gendered questions about her feelings, motherhood, and advice for young women. These questions played into the stereotypes of men and women in the fields of science and engineering; men were much more tough, rigorous, rational, impersonal, and unemotional, whereas women were more delicate, emotional, and nurturing. Throughout the press conference Ride was jovial, even though she faced some uncomfortable questions. She refused to answer a question about whether or not she had plans to be the first mother to travel in space. When asked how she adapted to working with men, Ride said, "It really hasn't taken any getting used to. I was a graduate student in physics before, and I was used to working with men in that field and women also." Had she been deferred to by the gentlemen on her crew? No, she responded, "In fact, Crip won't even open doors for me anymore." "I learned that right away," Crippen added. Ride often found the gender-biased questions overwhelming and said of the attention, "It's too bad our society isn't further along."[70]

A *Time* magazine reporter hoped to learn if Ride was the stereotypical hysterical woman who fell apart the moment something went wrong. He asked how she reacted when something did not go well in a training simulation—did she cry? Pilot Frederick H. Hauck, then sitting to Ride's right, recalled "she kind of gave him this 'You gotta be kiddin' me' kind of look and said, 'Why doesn't anyone ever ask Rick those questions?'"[71] Crippen jumped in and kidded, "the commander weeps."[72]

After the press conference reporters had lingering questions about whether or not Ride would wear makeup in space and what hygiene products she would use in space. Three days before launch Francine Brischel of the *Los Angeles Herald Examiner* inquired about Ride's makeup/personal hygiene/preference kit. What items did she decide to take with her? Ride took all the standard items in her personal hygiene kit, provided for all astronauts, except a razor. She took only a few items from the standard female preference kit: a head band, pony tail holder, tampons, and scissors. She chose not to take the kit's cosmetics. No one bothered to ask what the men took with them.[73] Once she was in orbit, the press reported that an American woman was finally in space without lipstick or any powder for those all important press conferences, videos, and still photographs.

Ride's flight was only one part of the story, however. Seven other women worked in the office in 1983. (Bonnie J. Dunbar and Mary L. Cleave were selected as part of the 1980 class of astronauts.) What did STS-7 mean to these women who had yet to fly onboard the space shuttle? It was one thing to select women, but the decision meant nothing until they actually flew in space. Ride's historic flight, Seddon hoped, would prove NASA's commitment to flying women onboard the shuttle and that women were "serious competitors" for shuttle flights, not just some public relations stunt for the space agency.[74]

Some people continued to believe that flying in space was a man's job, but her mission would—the women hoped—topple those feelings. STS-7, Sullivan explained, would "bring a very substantial demonstration, beyond all question that it (the Shuttle) works just as well with women aboard." Nothing bad would happen because a woman was onboard. "Keppler's [sic] laws are not spontaneously violated, the spacecraft will not fall from the sky. Believe it or not," she added, "the instrument panel does not know if it is a male or female hand that flips the switches."[75]

Resnik became the second woman in her class to be named to a flight, along with classmates Michael L. Coats, Steven A. Hawley, and Richard M. Mullane.

Henry W. Hartsfield commanded this crew of rookies known as the Zoo Crew. McDonnell Douglas employee Charles D. Walker later joined the crew as payload specialist for the continuous flow electrophoresis system (CFES). At the preflight press conference, Walker, not Resnik, received more questions from the press about the CFES device he would operate in space. Most of the questions she received focused on her specific role as a mission specialist. One reporter, however, did ask if Resnik intended to wear makeup or shave in space, as if applying makeup or shaving her legs in space was the most pressing concern she had on her first flight. She refused to say, believing that as a professional astronaut her technical work was subject to scrutiny, but not her personal preference or feminine hygiene kits. All astronauts (male and female), in her opinion, needed some modicum of privacy. Just because she was a woman did not warrant these types of questions.[76]

During her seven-year career Resnik chose to keep her personal life private, and as a result the articles published about her from 1978 forward tended to reflect the image she chose to project, instead of the feminine images some of the other women had to battle (Figure 6). The press occasionally mentioned her weight, her voice, demeanor, and her looks—"a pilot, with the looks of a stewardess, and the manner of a scientist"—but most of the articles focused upon her work or career.[77] She provided the basic facts about her life to reporters

FIGURE 6. Judy Resnik, in the backseat of a T-38 jet, talks with NASA pilot Richard Laidley. NASA Image S78-29248, courtesy of NASA.

in her NASA biographical data sheet, which often mirrored the details she discussed with the media, and in the few interviews she gave until her death in 1986. She chose not to discuss topics the media preferred to cover, so sometimes the press turned to her family for information. Recognizing that she was now in the public eye, Resnik preferred not to share those details. Flying onboard the shuttle was serious business, and she disliked answering what she deemed silly questions.[78]

Following Ride was a blessing for Resnik because she did not have to deal with the paparazzi as the second American woman astronaut. Plus, there were other firsts on her flight: the maiden flight of the *Discovery* orbiter, the first commercial payload specialist, and the first Frisbee-style deployment of a Leasat-2 satellite, which took some of the attention away from the fact that she was the second woman to fly. She could not, however, escape the fact that she was a woman and looked different from the men on her flight.

Two flight malfunctions forced Resnik to confront the very real differences between women and men astronauts, which she wanted to avoid. First, her long hair accidentally got caught in the IMAX camera's motor. When Hartsfield planned to call the ground and tell them about the problem, she was adamant that Mission Control not hear that her hair had caused the glitch. The men on the crew could not understand her annoyance. "What's the big deal?" they asked. Resnik, being second, realized that this incident demonstrated what the media had been saying for years—women and men astronauts were different—and she did not want them to continue telling this story. Eventually, the rest of the crew came to understand that no matter how successful Resnik was, the media would make the story about her sex, and she made them promise never to tell anyone what occurred onboard *Discovery*.[79] Later, the shuttle's waste collection system failed, and the crew had to use bags to capture their waste. Resnik was given the opportunity to use the toilet, but she refused. As Mullane explained, "Her use of the urinal would be a shout from the rooftops that a penis *was* necessary to deal with certain Shuttle emergencies."[80]

Interestingly enough, women's hair did come up at the postflight press conference. No one had told the press about the IMAX camera incident, but a reporter from Huntsville, Alabama, wanted to know if she had any suggestions about hairstyles for future women flying in space. She replied curtly, "No advice."[81]

Sullivan became the third American woman to fly in space and the first American woman to walk in space. Press interest in this flight was significantly less than the flight of STS-7, but there was interest because of several firsts: Sullivan's spacewalk and the first flight of two women on a space shuttle (Figure 7).

FIGURE 7. Kathy Sullivan and Sally Ride synchronize their watches before liftoff of STS-41G. NASA Image 41G-90082, courtesy of NASA.

Newspaper reporters wondered if women had the ability to complete a grueling spacewalk. Sullivan recalled that at the press conferences reporters would ask Crippen or Ride, who were also assigned to STS-41G, "Do you think Kathy can do this," instead of asking her directly.[82] Sullivan asserted she had the stamina and strength to do so. Her frame was similar to that of her spacewalking partner David C. Leestma, who fit into the same size spacesuit as Sullivan. As she explained, "I'm not your basic weak, frail female, and never have been. It's nice to find a place where qualifying for the [LA] Rams Front Four is an asset rather than a liability."[83] She successfully completed the spacewalk, proving that women were capable of the task. Coined a "giant step for womankind," the National Organization for Women equated her accomplishment with other key advancements that year, such as the selection of Geraldine Ferraro as the vice presidential candidate for the Democratic ticket.[84]

Interest in the flight of two women on STS-41G was mild; the celebrity of the flight seemed to be Canadian payload specialist Marc Garneau. Sullivan remembered waiting at the Ellington Field hangar with Ride to talk with Garneau, then taxiing in on another Gulfstream aircraft. When that plane arrived, the media crowded around the fuselage and staircase. Ride and Sullivan were stunned. "We joked that here we were, the first American woman in space, first American woman to fly twice, first American woman to walk in space, and a sea of media who appeared to not care."[85]

Media interest ramped up again for Fisher's first and only flight, STS-51A, in 1984, primarily because she was going to be the first mother in space. Coverage of the flight, a satellite deploy and retrieval mission, and the five-member crew varied. Biographical sketches printed in the *New York Times* identified the names of spouses and the number of children for each crew member. Fisher's profile, about twice as long as the rest of the crew, explained how she, a working mother, juggled career and family, but none of the male astronauts explained how they did so.[86] A majority of the newspaper photos of her featured a beaming mother with her daughter, either in the simulator or at KSC in their blue flight suits. The other photos tended to feature the two spacewalkers, Joseph P. Allen or Dale A. Gardner, riding the manned maneuvering unit or working in the payload bay of the shuttle to rescue two malfunctioning satellites.

The final two women to fly, Seddon and Lucid, received less interest than those who had flown in the previous two years. Journalists, preoccupied with covering the historic firsts of the shuttle program, found other crew members of interest as NASA added payload specialists to the crews. Seddon's flight included Utah Senator Jake Garn, and of the crew he received the bulk of the media's attention. Gone were the questions of "Do you think women can do this job" or "How's it different for women?"[87] During the flight the media turned their attention to the crew's efforts to activate a malfunctioning SYNCOM satellite. Lucid, the final woman to fly from her class, flew with an international crew that included a payload specialist from Saudi Arabia and another from France. *People* featured an article about Sultan Salman Abdulaziz Al-Saud, not Lucid, in its June 24, 1985, issue.[88] With the exception of a few hometown articles about Lucid, interest in STS-51G focused on the payload crew and its mission.

By the summer of 1985, all six had flown onboard the space shuttle at least once and were assigned to another mission. In January 1986, Resnik died along with six others when the *Challenger* broke apart during launch. President Ronald Reagan appointed a commission to investigate the disaster, and Ride was asked to serve on the panel. For the next two and a half years the accident, investigation, recovery, and return to flight of the shuttle program made headlines. Interest in writing about the women astronauts declined precipitously as flying women into space grew routine. As the media tended to hype firsts, the first six story faded. In 1987 Ride left NASA, leaving only four of the six women originally selected from the 1978 class. Two years later Fisher took a six-year leave of absence to raise her daughters and never flew again. Seddon and Sullivan each went on to fly two more missions, and Lucid completed a total of four flights and set endurance records on the Russian space station Mir.

The payloads on the women's second missions garnered more press than their inclusion on these flights. STS-34, which included Lucid, set the Galileo spacecraft on its path to Jupiter. Sullivan flew onboard STS-31, which deployed the Hubble Space Telescope. STS-40, also known as Spacelab Life Sciences 1 (SLS-1), was dedicated to the study of life sciences and featured, for the first time, three women, including Seddon. When compared with other stories about SLS-1, the flight of three women was merely a footnote.

The decision to select six women astronaut candidates for the shuttle program was hardly minor, however, as illustrated by the media interest. It is clear that the media covered the women differently because of their sex, a fact that the women recognized. "We do the same work as our male colleagues, yet nobody asks them about their families, or how their kids feel about their work," Lucid complained. Being an astronaut required a certain set of skills, a science or medical degree, but "that doesn't have anything to do with being a woman or a wife or a mother."[89] Over time that changed, not because the media became less biased in its coverage of women, but because other NASA people, payloads, and space program events demanded the attention of journalists and writers. Nonetheless, these six proved that they were more than mothers, wives, or pretty women. Their achievements in space and on Earth spoke volumes about women's abilities, which matched or exceeded any male astronauts' accomplishments on the ground and in zero gravity.

Notes

1. Kathryn D. Sullivan, interview by author, May 10, 2007, transcript, JSC Oral History Project, JSC History Collection, University of Houston-Clear Lake, Houston, TX (hereafter cited as JSC OHP).

2. Amy E. Foster contended that the media interest "likely surpassed all but the coverage given to the original Mercury Seven astronauts and the historic Apollo 11 crew." *Integrating Women into the Astronaut Corps: Politics and Logistics at NASA, 1972–2004* (Baltimore: Johns Hopkins University Press, 2011), 95.

3. Kraft said that her family had not been taken into consideration. "Space Shuttle Astronauts Press Conference," January 16, 1978, NASA Headquarters Historical Reference Collection, https://mira.hq.nasa.gov/history/ (accessed October 17, 2011).

4. Barry Hart, "The Glamornauts," *Globe*, March 18, 1980, Anna Fisher Biographical File, Record No. 661, NASA Headquarters Historical Reference Collection, Washington, DC (hereafter cited as HQ Collection); Don Retson, "America's Space Queens," *National Examiner*, August 4, 1981, 20–21.

5. The media's paradoxical messages about women are covered in several important works. Susan J. Douglas, *Where the Girls Are: Growing Up Female with the Mass Media* (New York: Times Books, 1995); Sherrie A. Inness, *Tough Girls: Women Warriors and Wonder Women in Popular Culture* (Philadelphia: University of Pennsylvania Press, 1999).

6. This list represents a sampling of the works that highlight these differences. Cynthia Carter and Linda Steiner, "Texts in Context," in *Critical Readings: Media and Gender*, ed. Cynthia Carter and Linda Steiner (Maidenhead, UK: Open University Press, 2004), 37–40; Rosalind Gill, *Gender and the Media* (Malden, MA: Polity Press, 2007), 113–20; Ray Jones, Audrey J. Murrell, and Jennifer Jackson, "Pretty Versus Powerful in the Sports Pages: Print Media Coverage of U.S. Women's Olympic Gold Medal Winning Teams," *Journal of Sport and Social Issues* 23 (May 1999): 183–92; Jessica Aubin, Michelle Haak, and Andrew Mangini, "Media Coverage of Women Candidates," *White House Studies* 5, no. 4 (October 2005): 523–37.

7. The focus on femininity and appearance in an election can harm a woman's chances. Nathan A. Heflick and Jamie L. Goldenberg, "Sarah Palin, a Nation Object(ifie)s: The Role of Appearance Focus in the 2008 U.S. Presidential Election," *Sex Roles* 65 (2011): 149–55; Caroline Heldman, Susan J. Carroll, and Stephanie Olson, "'She Brought Only a Skirt': Print Media Coverage of Elizabeth Dole's Bid for the Republican Presidential Nomination," *Political Communication* 22 (2005): 315–35.

8. Marjorie Ferguson, *Forever Feminine: Women's Magazines and the Cult of Femininity* (London: Heinemann, 1983).

9. Inness, *Tough Girls*, 56–57.

10. Junetta Davis, "Sexist Bias in Eight Newspapers," *Journalism Quarterly* 59, no. 3 (Autumn 1982): 456–60.

11. Roy E. Blackwood, "The Content of News Photos: Roles Portrayed by Men and Women," *Journalism Quarterly* 60, no. 4 (Winter 1983): 714.

12. W. James Potter, "Gender Representation in Elite Newspapers," *Journalism Quarterly* 62, no. 3 (Autumn 1985): 636–40.

13. Rita J. Simon and Jean M. Landis, "The Polls—A Report: Women's and Men's Attitudes about a Woman's Place and Role," *Public Opinion Quarterly* 53 (1989): 268.

14. Aubin et al., "Media Coverage of Women Candidates," 530.

15. Susan Faludi, *Backlash: The Undeclared War against American Women* (New York: Anchor Books, 1992).

16. Naomi Wolf, *The Beauty Myth: How Images of Beauty Are Used against Women* (New York: William Morrow and Company, 1991), 10.

17. For examples, see Sharon Herbaugh, "Female Astronauts They Can't Wait to Space Out," *Dallas Morning News*, May 11, 1980; Susan Witty, "Our First Women in Space," *Geo*, September 1982, 21.

18. Leslie Williams, "NASA Space Cadets Score Several Firsts," *Bay Saint Louis (MS) Sea Coast Echo*, December 3, 1978, Fisher Biographical File.

19. Rhea Seddon, "The Lady Astronaut's Diet," *Weight Watchers*, May 1979, 28, Rhea Seddon Biographical File, Record No. 2939, HQ Collection.

20. Sullivan, interview.

21. "Inside NASA," *Washington Post*, October 1, 1984, Kathryn D. Sullivan Biographical File, Record No. 2953, HQ Collection.

22. "NASA Picks Six Women Astronauts with the Message: You're Going a Long Way Baby," *People Weekly*, February 6, 1978, 28–30, 35.

23. Margaret Carlisle Duncan, "The Politics of Women's Body Images and Practices: Foucault, the Panopticon, and Shape Magazine," *Journal of Sport and Social Issues* 18 (February 1994): 51.

24. Wolf, *The Beauty Myth*, 200.

25. Andrea Fooner, "Countdown to Fitness: How an Astronaut Stays in Shape," *Redbook: The Magazine for Young Women*, June 1978, 112.

26. Anna L. Fisher, interview by author, February 17, 2009, transcript, JSC OHP.

27. Fooner, "Countdown to Fitness," 113.

28. Fooner, "Countdown to Fitness," 172, 174.

29. Seddon, "The Lady Astronaut's Diet," 28.

30. Below the article's title *Redbook* highlighted the name and department that produced the piece.

31. Other popular magazines reinforced the idea that women would not develop muscles if they exercised with weights in the 1970s and 1980s. "Exercise with an Astronaut," *Redbook*, January 1980,

89–91; Pirkko Markula, "Firm but Shapely, Fit but Sexy, Strong but Thin: The Postmodern Aerobicizing Female Bodies," *Sociology of Sport Journal* 12 (1995): 431–32.

32. Caroline Seebohm, "Shaping up for Space: America's First Women Astronauts—How They Were Chosen, Plus the Health and Fitness Program That Will Help Them Feel at Home in the Stars," *House and Garden*, July 1978, 168.

33. "NASA Picks Six Women," 28–29.

34. Paul Recer, "Enter the Specialist—A New Breed of Astronaut," *U.S. News and World Report*, February 23, 1981, 60, 62.

35. Kent Biffle, "Woman 'Goes into Orbit' over Job," *Dallas Morning News*, January 25, 1978.

36. Ferguson, *Forever Feminine*, 117.

37. David Fritze, "New Heights, Space Flights," *Oklahoma Monthly*, March 1978, 70.

38. Simon and Landis, "The Polls," 269.

39. Sara Terry, "New Astronaut Set for Move," *Christian Science Monitor*, April 17, 1978, Fisher Biographical File.

40. "NASA Picks Six Women," 28–29.

41. Ferguson, *Forever Feminine*, 109–110, 189.

42. Susan J. Douglas and Meredith W. Michaels, *The Mommy Myth: The Idealization of Motherhood and How It Has Undermined Women* (New York: Free Press, 2004), 1–27, 138.

43. The article, however, focused more upon her training and chemistry skills than other essays. "Woman Biochemist Is Astronaut Candidate," *Chemical and Engineering News*, May 1, 1978, 21–22.

44. See, for instance, Fritze, "New Heights, Space Flights," 67; Philip Finn, "Dream Comes True for First Mom Chosen to Make Historic Space Flight," *Star American Women's Weekly*, February 7, 1978; "NASA Picks Six Women," 30.

45. Vivian Vahlberg, "City Woman Is Selected as Astronaut," *Oklahoma City Times*, January 16, 1978.

46. This theme continued to be a subject of interest almost a decade after Ride's historic flight. Judy Stein, "Six Who'll Dare," *Texas Woman Magazine*, January 1980, 59; Marjorie Miller, "Mothers in Danger," *Working Woman*, May 1993, 70–73, 98.

47. John Lawrence, "Expectant Astronauts," Release No. 82-025, April 28, 1982, JSC Public Affairs Office, Houston, TX.

48. See, for instance, "Expectant Astronaut," *Alton (IL) Telegraph*, April 30, 1982; "Astronaut to Take Maternity Leave," *Lawrence (KS) Journal World*, April 30, 1982.

49. Rhea Seddon, interview by author, May 20, 2010, transcript, JSC OHP.

50. See, for instance, "Astronaut Gives Birth," *Winchester (WV) Star*, August 11, 1983.

51. Fisher, interview.

52. "Here's Looking at You Kids," *People Weekly*, December 27, 1982, 156.

53. "Meet the First Mothers to Fly in Space," *Good Housekeeping*, January 1984, 122–23.

54. "2 Astronaut Moms Picked for Flights," *USA Today*, September 22, 1983, Fisher Biographical File.

55. For more information on this theme, see chapter 3 of Douglas and Michaels, *The Mommy Myth*.

56. Aimee Lee Ball, "When Mom Is an Astronaut," *Parade Magazine*, October 28, 1984, 6–7.

57. John M. Fabian, interview by author, February 10, 2006, transcript, JSC OHP.

58. Retson, "America's Space Queens," 21.

59. Dustin Harp, Jaime Loke, and Ingrid Bachmann, "First Impressions of Sarah Palin: Pit Bulls, Politics, Gender Performance, and a Discursive Media (Re)contextualization," *Communication, Culture, and Critique* 3 (2010): 291–309.

60. Jeanne Forbis, "Space Spirited," *Tulsa World*, September 21, 1980.

61. Retson, "America's Space Queens," 20.

62. Biffle, "Woman 'Goes Into Orbit.'"

63. Ball, "When Mom Is an Astronaut," 5–6.

64. Dan Weisman, "Kathy Sullivan, an Astronaut and a Warm, Friendly Person," *Daily Citizen*, September 29, 1979, Sullivan Biographical File; Sullivan, interview.

65. Lawrence, "Three Shuttle Crews Announced," Release No. 82-023, April 19, 1982, JSC Public Affairs Office, Houston, Texas.

66. Kathleen Hendrix, "First Female Astronaut Maintains Status as a Very Private Person," *Houston Chronicle*, May 17, 1982.

67. STS-7 and STS-8 Prime Crew Press Conference, April 29, 1982, DVD, JSC Media Resources Center, Houston, TX.

68. Hendrix, "First Female Astronaut."

69. Susan Okie, "The Sally Ride Story," *Today*, May 8, 1983.

70. STS-7 Preflight Press Conference, May 24, 1983, DVD, JSC Media Resources Center, Houston, TX.

71. STS-7 Preflight Press Conference; Frederick H. Hauck, interview by author, November 20, 2003, transcript, JSC OHP.

72. STS-7 Preflight Press Conference.

73. STS-7 Traditional Query Book, JSC Public Affairs Office, Houston, TX.

74. Olive Talley, "Flight of Fancy for America, but a Day's Work for Ride," *Washington Times*, June 14, 1983, Sally Ride Biographical File, Record No. 1795, HQ Collection.

75. "7 Women Astronauts Hope for Ride's Success," *Houston Post*, June 20, 1983.

76. The women disliked questions about makeup and emotions. Jim Asker, "2nd Woman in Space Just Glad to Be There," *Houston Post*, September 2, 1984; "When Shuttle Astronauts Meet the Press," *Space World*, September 1985, 5.

77. Carrie Dolan, "Success Stories," *Wall Street Journal*, June 29, 1983.

78. Staff of the *Washington Post*, *Challengers: The Inspiring Life Stories of the Seven Brave Astronauts of Shuttle Mission 51-L* (New York: Pocket Books, 1986), 81–83, 88, 94.

79. Richard M. Mullane, interview by Rebecca Wright, January 24, 2003, transcript, JSC OHP; Steven A. Hawley, interview by Sandra Johnson, December 4, 2002, transcript, JSC OHP.

80. Italics in original. Mike Mullane, *Riding Rockets: The Outrageous Tales of a Space Shuttle Astronaut* (New York: Scribner, 2006), 183.

81. STS-41D Postflight Press Conference, September 12, 1984, STS-41D Documents, Box 3, Shuttle Series, JSC History Collection, University of Houston-Clear Lake Archives, Houston, TX.

82. Sullivan, interview.

83. Jim Asker, "Astronaut Sullivan Thrives on Hard Work," *Houston Post*, October 11, 1984.

84. Jim Asker, "Sullivan Takes Giant Step for Womankind," *Houston Post*, October 12, 1984; "Briefing," *New York Times*, December 26, 1984, Sullivan Biographical File.

85. Kathryn D. Sullivan, interview by author, March 12, 2008, transcript, JSC OHP.

86. "Sketches of Five Astronauts on Mission," *New York Times*, November 9, 1984, Fisher Biographical File.

87. Rhea Seddon, interview by author, May 21, 2010, transcript, JSC OHP.

88. Montgomery Brower, "Saudi Prince Salman Al-Saud Gets Set for a Heavenly Tour of Allah's Creation," *People Weekly*, June 24, 1985, 34–35.

89. Janis Williams, "Make Way for the Ladies in Space," *The Saturday Evening Post*, September 1982, 45.

CHAPTER 9

Warriors and Worriers:
Risk, Masculinity, and the Anxiety of Individuality
in the Literature of American Spaceflight

Margaret Lazarus Dean

Spaceflight in literature has generally been relegated to science fiction, a genre based on the power of speculation. Even when a work of science fiction is grounded to some degree in historical reality, the speculative element is always key to the narrative.[1] But a number of works of literature have taken on historically accurate depictions of spaceflight in ways that could not be classified as speculative; these books and stories are more aptly classified as a subcategory of historical fiction, one that takes a specific aspect of American history as their setting.

Alessandro Manzoni, a poet and historical novelist himself, defines the difference between history and historical fiction as follows:

> If one takes away from the poet what distinguishes him from a historian, the right to invent facts, what is left? Poetry; yes, poetry. For what, in the end, does history give us? Events that are known only, so to speak, from the outside, what men have done. But what they have thought, the feelings that have accompanied their decisions and their plans, the words by which they have asserted—or tried to assert—their passions and wills on those of others, by which, in a word, they have revealed their individuality: all that, more or less, is passed over in silence by history: and all that is the domain of poetry.[2]

Some historians might take issue with the limitations Manzoni ascribes to their work, but writers of literature do bear a special burden to create an entertaining story that makes emotional and thematic sense, that includes characters, emotions, theme, and point of view. A work of history has to answer the question "What exactly happened?" and, to some extent, "What does it mean that this happened?" But to put a finer point on Manzoni, a novel or other work of fiction must also answer the question "How must it have felt to have been the person from whose point of view this story is being told?" This issue of point of view is key to the study of fiction, as it goes to the heart of not only what events and situations we see in the narrative but also the inflection with which those events and situations are told, which is to say the meanings they are given. Examining the ways in which historic events are inflected through fictional characters' points of view can lead us to a deeper understanding of the cultural and emotional meanings of spaceflight.

The first era of spaceflight, comprising the Mercury, Gemini, and Apollo projects (sometimes called the "heroic era"), gave rise to a wide range of fictional and semifictional representations, including Tom Wolfe's *The Right Stuff*, Norman Mailer's *Of a Fire on the Moon*, and James Michener's *Space*.[3] The space shuttle era, in contrast, has produced noticeably fewer books and films, and those fictional narratives that do use spaceflight as their setting tend to include nonastronaut characters as main characters alongside, or instead of, astronauts. It is worth noting that the most popular representations of spaceflight to appear in this era (for instance, the film *Apollo 13* [1995] and the miniseries *From the Earth to the Moon* [1998]) do not tell stories of contemporaneous spaceflight but instead reach back to the heroic era for their settings.[4]

This chapter will examine the representation of spaceflight and spacefarers in two popular books, one from the sixties era of spaceflight and one from the space shuttle era. In the sixties era, Tom Wolfe's bestseller *The Right Stuff* captured the imagination of readers and helped define the image of the astronaut for a generation. Many of the emotional and cultural currents we have come to associate with the heroic era are brought to life in Wolfe's interpretation of the Mercury astronauts' experiences.

The space shuttle era produced no works of literature nearly as well known as *The Right Stuff*, and indeed, the very lack of a wildly popular narrative about shuttle astronauts illustrates some of the key differences between the two eras. In the space shuttle era, astronauts became more numerous and therefore more anonymous, their ranks infiltrated by women, minorities, and nonpilot mission specialists. The reusability and apparent safety of the space shuttle (as compared to sixties era spacecraft) and the relative closeness of low Earth orbit (as compared to the Moon) also change the popular perception of the astronauts, whose identities easily become bound up with their vehicles and their destinations. The space shuttle era novel that comes closest to being an analogue to *The Right Stuff* is Stephen Harrigan's 2006 novel *Challenger Park*, in which an astronaut is one of two main characters and a space shuttle mission forms the main plot arc.[5] Although *Challenger Park* did not become part of the popular imagination in the way *The Right Stuff* did, it does reflect the cultural uses to which astronauts and other characters are put, the roles they play in our imagination of what spaceflight means in the shuttle era. Comparing the two books reveals a distinct shift in the popular imagination of what spaceflight means, including significant differences in the roles of astronauts, astronauts' spouses, and nonastronaut space workers. Specifically, this comparison will reveal a number of common themes and anxieties revolving around the intertwining binaries of risk versus safety, masculinity versus femininity, and individuality versus collectivity. As we will see, the space shuttle era will merge and complicate the meanings of roles that had been simple and distinct in the sixties era of spaceflight.

The Sixties Era and *The Right Stuff*

Tom Wolfe's *The Right Stuff*, published in 1979, chronicles Project Mercury through the points of view of the seven test pilots who would become the first astronauts: their selection, training, flights, and return to Earth.[6] Written by a novelist, the book was based on true events and can be categorized, using a more contemporary term, as creative nonfiction, a genre that applies the tactics of literary fiction writing to nonfiction accounts. What makes the book a creative rather than historical or biographical work is Wolfe's willingness to reinvent scenes and dialogue with a level of detail he cannot have known from the point of view of his characters, to invent thoughts and emotions in ways that historians or journalists never would. As a result, the themes in *The Right Stuff* reflect Wolfe's *imagination* of spaceflight and of spacefarers nearly as much as would a completely fabricated work of fiction. In a book often regarded as one of the most important in establishing the new genre of creative nonfiction, Tom Wolfe, a writer already famous for his brash, colorful style, turned his attentions to a subject as big as his voice.

It is important to note the publication date of *The Right Stuff*. Although it covers events from the late forties to the early sixties, the book was researched and written with a significant amount of hindsight, during the quiet stretch between Apollo 17, the last lunar landing, and the first space shuttle launch in April 1981. Indeed, Wolfe's foreword notes that he first got the idea for the project when he met several former astronauts at the launch of Apollo 17 in December 1972.[7] Wolfe goes on to describe his motivation for writing the book: to understand what gave the astronauts the courage to undertake such daring missions. "What is it, I wondered, that makes a man willing to sit up on top of an enormous Roman candle . . . and wait for someone to light the fuse?"[8] The answer, as Wolfe constructs it, is a hard-to-define set of characteristics, the so-called Right Stuff, a definition that both confirms and complicates common assumptions about what motivated astronauts and about what Americans valued most in a hero in that era.⁹

A first attempt at a definition of the Right Stuff comes early in the book:

> The world was divided into those who had it and those who did not. This quality, this *it*, was never named, however, nor was it talked about in any way.
>
> As to just what this ineffable quality was . . . well, it obviously involved bravery. But it was not bravery in the simple sense of being willing to risk your life . . . No, the idea here (in the all-enclosing fraternity) seemed to be that a man should have the ability to go up in a hurtling piece of machinery and put his hide on the line and then have the moxie, the reflexes, the experience, the coolness, to pull it back in the last yawning moment—and then to go up again *the next day*, and the next day, and every next day, even if the series should prove infinite.¹⁰

The fact that the astronauts were drawn from a pool of test pilots, nearly all of them veteran fighter pilots, had specific effects on the culture of astronauts having to do with competitiveness and domination. Each of the men had already withstood a series of tests meant to weed out the weak at each level, and (according to Wolfe) each of them had come to see the world as a never-ending series of tests of manhood:

> A career in flying was like climbing one of those ancient Babylonian pyramids made up of a dizzy progression of steps and ledges, a ziggurat, a pyramid extraordinarily high and steep; and the idea was to prove at every foot of the way up that pyramid that you were one of the elected and anointed ones who had the right stuff and could move higher and higher and even—ultimately, God willing, one day—that you might be able to join that special few at the very top, that elite who had the capacity to bring tears to men's eyes, the very Brotherhood of the Right Stuff itself.¹¹

In this passage, and in many others like it throughout the book (Wolfe returns to the ziggurat metaphor repeatedly), the Right Stuff is constructed as being not only a set of characteristics a man might possess within himself but also a competition by which a man can measure himself against other men. It is not enough for a pilot to land a jet on an aircraft carrier at night; he must do so more times, and on choppier seas, than other men. It is not enough for a man to sit atop an experimental rocket and wait for someone to light the fuse; he must do so with a heart rate lower than those of the men who preceded him. And so on. In a critical essay on *The Right Stuff*, Brian Abel Ragen points out that Wolfe's previous projects had been on Ken Kesey's Merry Pranksters and the Hell's Angels: "[Wolfe] is dealing with his constant theme of status, only here status is achieved not through purely symbolic things like clothing but by skill and action."[12]

But a contradiction lurks within the definition of the astronauts' heroism: the seven Mercury astronauts were the best pilots of their generation, but the Mercury spacecraft itself required no piloting. Tom Wolfe sums up the contemptuous attitude among pilots still flying experimental test planes at Edwards Air Force Base, among them the legendary Chuck Yeager: "The astronaut would not be expected to *do* anything; he only had to be able to take it."[13] The astronauts had been chosen as the best of their kind, implying that they were at the top of the ziggurat, yet without the chance to prove themselves by *controlling* the spacecraft, they ran the risk of losing their masculinity or even their humanity.[14]

This anxiety was expressed by constant jokes and reminders that "a monkey's gonna make the first flight."[15] To men who take pride in proving themselves, flying a mission that, apparently, can be performed equally well by a chimpanzee was disconcerting. Similarly, a group of female pilots who went through medical testing at Lovelace Clinic in Albuquerque, New Mexico and hoped to be considered as astronauts came up against the cultural mandate to define the astronaut as a role that only a man could play. As Margaret Weitekamp points out in *Right Stuff, Wrong Sex*, the common assumption was that

> if a woman could fly a craft, it must be easy to do. Indeed, Western space experts cited [female Soviet cosmonaut Valentina] Tereshkova's mission as evidence that anyone could occupy the automated *Vostok* capsules. In contrast, the arguments went, America's Mercury spacecraft required skilled astronaut pilots with engineering backgrounds and experience in test flying military jets.[16]

What made our spacecraft superior, in other words, was the fact that it took a man to fly it; therefore, allowing a woman to fly would undermine the entire program. The contradiction inherent in maintaining this argument while also preparing to send chimps for the first flights was not lost on the Lovelace women. As pilot Jerrie Cobb noted in a 1962 congressional hearing on whether women should be considered,

> I find it a little ridiculous when I read in a newspaper that there is a place called Chimp College in New Mexico where they are training 50 chimpanzees for space flight, one a female named Glenda. I think it would be at least as important to let the women undergo this training for space flight.[17]

The well-publicized first flight by Ham the chimp in January 1961 underscored the fact that the Mercury astronauts would not actually be called upon to perform as pilots, that they need only be able to withstand the rigors of spaceflight, to "take it," while following orders from the ground. This reality undermined the masculinity inherent in pilot lore and created the unflattering impression that the so-called Right Stuff was actually something along the lines of passivity and obedience. In fact, suggestions that the Lovelace women might make *better* astronauts because they were smaller than the men, consumed fewer resources, and outperformed the men on some tests further called into question the masculinity of the Mercury 7 in ways similar to the chimpanzees' flights. In shoring up the requirement that an astronaut must be a man, Wolfe's narrative reveals, the Mercury 7 were threatened by two separate subdefinitions of the word: a man is the opposite of a woman, and he is also the opposite of a monkey.

This set of contradictions is very much at work in Wolfe's descriptions of the potential astronauts' testing at the Lovelace Clinic. Although the testing was, in one sense, yet another chance for the pilots to compete with each other and climb the ziggurat, it was also a series of exercises in discomfort, pain, and humiliation. In Wolfe's retelling, the pilots were incensed by the intrusiveness of the examinations and at the presumptuousness of the doctors, nurses, and psychologists, who treated them as test subjects rather than as officers and heroes.

> At Lovelace, in the testing for Project Mercury, the natural order was turned upside down. These people not only did not treat them as righteous pilots, they did not treat them as pilots of any sort, they never even alluded to the fact that they were pilots. An irksome thought was beginning to intrude. In the competition for *astronaut* the kind of stuff you were made of as a *pilot* didn't count for a goddamned thing.[18]

The fact that the personnel at Lovelace failed to treat the men with the respect to which they had become accustomed rankled—so much so that, after losing a showdown with a female psychologist, Pete Conrad lost his temper and confronted a superior officer about the indignity of their treatment. He was later convinced that this outburst cost him a place among the Mercury 7 (he was chosen among the Next Nine and subsequently flew on Gemini V, Apollo 12, and Skylab 2).[19] In Wolfe's retelling, Conrad's rebellion was not portrayed as an example of childish petulance or as a lack of emotional control: it was a courageous stand for the officers' dignity, another manifestation of the Right Stuff. He was willing to risk a chance at spaceflight to stand up for his and his brothers' right to be treated as heroes, not as physical bodies. It is interesting to note that some of the procedures Wolfe details are similar to common practices in obstetrics in the late fifties and early sixties; one wonders whether the Lovelace women, some of whom had given birth, were as outraged by the tests or the treatment they received. Unlike men, women were expected to accept routine poking, prodding, and physical humiliation without complaint. Of course, this is precisely why the tests were so threatening to the masculinity of the Mercury 7.

Gender and Division of Labor in *The Right Stuff*

The use of the words "fraternity" and "brotherhood" in the book's first definitions of the Right Stuff is no accident. As Tom Wolfe portrays them, the community of test pilots from which the first astronauts were drawn was a culture deeply rooted in values of masculinity. The burden of risking life and limb to defend one's country was understood to come along with a set of male privileges—drinking and driving and womanizing among them. These behaviors were not considered embarrassments or potential scandals to be reigned in; as Margaret Weitekamp points out, "Such macho excesses did not worry NASA decision makers. The space agency viewed this particular kind of manhood as part and parcel of the talents NASA needed."[20]

Given the intense focus on masculinity in Tom Wolfe's book (and, as Weitekamp's and others' research reveals, the vigor with which the all-male astronaut corps was defended), it is interesting to note that the opening scene of *The Right Stuff* takes as its point-of-view character a woman. Jane Conrad is a test pilot's wife at Jacksonville Naval Air Station in 1955, and the opening scene finds her frantically phoning other wives when a rumor spreads that "something has happened out there."[21] As the hours go by and fewer and fewer pilots are left unaccounted for, Jane waits to learn her husband's fate (bad news must be delivered in person and by a male figure of authority, ideally a clergyman—never by another woman).[22] When it turns out that Jane's husband Pete is unharmed and a close friend of his has been killed, Jane still must attend a funeral and comfort a widow, soon to be followed by another, then another.

Throughout the book, the astronauts' wives are important figures; several of them become point-of-view characters for sequences detailing their thoughts and feelings. Each time we see the inner life of one of the wives, she betrays emotions that the astronauts themselves can never admit to having—chiefly, fear of the risks involved in flight test and space travel. In the first press conference the Mercury astronauts hold in 1959, the astronauts are surprised (again, according to Wolfe) that most of the questions had to do with their wives, children, and churches rather than with their exploits as pilots. Wolfe reflects,

> All the questions about wives and children and faith and God and motivation and the Flag . . . they were really questions about

widows and orphans . . . and how a warrior talks himself into going on a mission in which he is bound to die.²³

Questions about wives are really questions about widows—in other words, about the astronauts' potential deaths. Once the wives are introduced to the media, they are often photographed and quoted alongside their husbands to give a face to what is a stake. The astronauts' roles as husbands and as risk takers become one and the same; if the Right Stuff calls for an astronaut to present a performance of fearlessness, risk cannot be properly understood as risk without a woman left behind to worry about him.

Notions of gender are completely intertwined with notions of risk, then, in *The Right Stuff*; the danger astronauts face is integral to our understanding of their rarefied abilities, yet the fear that makes that risk real can only be expressed by the women. Wives are put to work constructing the Right Stuff in various ways, both subtle and unsubtle. The physical daring and acumen possessed by the astronauts is always constructed as a uniquely masculine trait, but in the culture of *The Right Stuff*, each masculine warrior needs a feminine worrier at home to make his heroism complete. It is no accident that no bachelor astronauts were chosen for Project Mercury.

The wives' roles as official worriers is nowhere as obvious as during the launches, when the press set up vigils on their lawns. Here Tom Wolfe describes the scrutiny Louise Shepard was subjected to while her husband became the first American in space:

> Before she knew it she was caught up in the same psychology that works at a *wake*. She was suddenly the central figure in a Wake for My Husband—in his hour of danger, however, rather than his hour of death. The secret of the wake for the dead was that it put the widow on stage, whether she liked it or not. In the very moment in which, if left alone, she might be crushed by grief, she was suddenly thrust into the role of hostess and star of the show. It's free! It's *open house*! Anybody can come on in and gawk! Of course, the widow can still turn on the waterworks—but it takes more nerve to do that in front of a great gawking mob than it does to be the brave little lady, serving the coffee and the cakes. For someone as dignified and strong as Louise Shepard, there was no question as to how it was going to come out. As hostess and the main character in this scene, what else was there for the pilot's wife to do but set about pulling everybody together?²⁴

It is the role of Mrs. Shepard, as an astronaut's wife, both to bring to life for us the danger her husband faces by playing the role of the worrier and also to "pull everybody together" even if the worst should happen. Whenever one of the test pilots is killed, the women are expected to hew to a prescribed narrative of those deaths; in a passage told from the point of view of the wives, Wolfe dramatizes their experience of sitting through their husbands' discussions of accidents that killed their comrades, discussions that always blamed the dead pilot for his own demise, always found a reason why the living would have survived the same disaster:

> Every wife wanted to cry out: "Well, my God! The machine broke! What makes any of you think you would have come out of it any better!" Yet intuitively Jane and the rest of them knew it wasn't right even to suggest that . . . It seemed not only wrong but dangerous to challenge a young pilot's confidence by posing the question.[25]

In order to keep going, the surviving pilots need to believe they would not have suffered the same fate. This is perhaps a useful psychological defense mechanism. But here their wives are expected to withhold their observations about the risks of their husbands' endeavors—that these accidents could in fact happen to any of them—an obvious contradiction from their other responsibility as designated worriers. The women are expected to perform their concern for the astronauts' safety, yet they also bear the responsibility of maintaining their husbands' confidence through deft handling of their delicate egos. It starts to seem as though the Right Stuff—the very courage that had seemed to be an inherent trait in the Mercury astronauts—is a commodity supplied to them in large part by complex performances of their wives.

Geeks and Individuality in *The Right Stuff*

Throughout *The Right Stuff*, Tom Wolfe defines the astronaut using the archaic concept of single-combat warfare, a form of battle in which the strongest warriors from each side fought to the death in lieu of entire armies slaughtering each other. The rewards for single-combat warrior status include the warrior's being lionized as a hero, not as a soldier among many, but as a unique man who possesses abilities the rest of us lack. Wolfe implies that the obsessive focus on the seven Mercury astronauts as individuals was constructed in direct opposition to the perceived faceless interchangeability of workers in the Soviet Union. Of course, Americans understood that the astronauts were not getting to space by themselves, but the thousands of engineers, scientists, managers, technicians, and other space workers were easy to ignore in favor of the single-combat warriors. This emphasis on the astronaut as triumphant individual is alive and well in *The Right Stuff*: the focus on astronauts in this book is so pervasive that it nearly excludes the roles played by anyone else in the accomplishments of the space program, a serious inaccuracy if not an uncommon one.

Here Wolfe points out that none of the astronauts' daring would be possible without the recent invention of electronic computers, but through humor he undercuts the possibility of a heroic role for the engineers:

> Engineers . . . were creating, with computers, systems in which machines could communicate with one another, make decisions, take action, all with tremendous speed and accuracy, . . . Oh, genius-engineers! Ah yes, there was such a thing as self-esteem among engineers. It may not have been as grandiose as that of fighter jocks . . . nevertheless, many was the steaming enchephalitic [sic] summertime Saturday night at Langley when some NASA engineer would start knocking back that good sweet Virginia A.B.C. Store bourbon on the patio and letting his ego out for a little romp, like a growling red dog.[26]

The actual engineers whose achievements in computer engineering made spaceflight possible might have been less than flattered by Wolfe's assurance that "there was such a thing as self-esteem among engineers," but the clear bias toward individual achievement over group effort brought to life here interacts intriguingly with the other themes at work in *The Right Stuff*. The anxiety about individuality, which at a fundamental level was an anxiety about communism, goes so deep that some of Wolfe's astronauts distrust members of their own ranks who happen to have an engineering background:

> They would be knocking back a few at somebody's house, some Saturday night, and they would hear [Gordon] Cooper starting to talk about something extraordinary that had happened when he was testing the F-106B or whatever at Edwards . . . and the blood would come into somebody's baleful eyes, and he'd say, "I'll tell you what Gordo did at Edwards. He was in *engineering*." The way engineering was pronounced, you would have thought Gordo had been a quartermaster or a drum major or a chaplain . . . To be in engineering was to be an also-ran.[27]

Incidentally, this fear of communism was also used to justify maintaining an all-male astronaut corps. Valentina Tereshkova's flight was interpreted in the West as a symbol of the way in which communism erased gender differences. The Soviets made their women work in factories and fly in space, stripping them of femininity, but American women were treated to more chivalrous treatment. Again, essentialism and idealism get mixed up: if Soviet women can fly successfully as cosmonauts, then there is nothing inherent in human females that *precludes* them from being able to take part in this aspect of society, yet this is exactly what the Lovelace women were told over and over.

Anxieties about the interchangeability of engineers, physicists, and other space workers carries through in telling ways in later novels about spaceflight. As fear of communism abates, collective work becomes less threatening, and nonastronaut space workers can be seen as individuals, main characters, and even heroes. Histories and nonfiction narratives about spaceflight moved away from the assumption that the single-combat warrior was at the center of the drama and began to feature the sorts of space workers that made Wolfe uncomfortable.[28] Roger Launius points out that the heroes of the shuttle era film *Apollo 13* (released in 1995) were not the astronauts but "the geeks of Mission Control" who solved technical problems to bring the astronauts home safely.[29] Similarly, the miniseries *From the Earth to the Moon*, produced by Tom Hanks after the success of *Apollo 13*, examined the lives of many nonastronaut space workers and helped to reinvent a new type of space hero for the space shuttle era.

The Space Shuttle Era and *Challenger Park*

In the space shuttle era (1981–2011), the meaning of astronaut as masculine conqueror and single-combat warrior began to erode. The astronaut class introduced in 1978 included scientists, women, minorities, and nonpilots, changing the definition of "astronaut" and the meaning of spaceflight.[30] Stephen Harrigan's *Challenger Park* takes as one of its two main characters Lucy Kincheloe, an astronaut living a mundane life in suburban Houston with her husband and two children in about 2002 (after 9/11 but before the *Columbia* disaster). She finds herself pulling away from her husband, also an astronaut, and starting an affair with her training team leader, Walt, at the same time her own first mission is drawing near. Half the chapters are told from Lucy's point of view and half from Walt's point of view as he agonizes over falling in love with Lucy and worries about her safety when she is traveling in space. This narrative is distinct from Mercury/Gemini/Apollo era narratives most saliently in Lucy's character: she is an astronaut's wife, a mother of two children, and a scientist, but in the shuttle era she is also an astronaut herself. Maybe just as remarkable, one of the two main characters is a nonastronaut space worker. Although "the geeks of Mission Control" became the heroes of *Apollo 13*, *Challenger Park* may be the first novel that offers one of the geeks a direct point of view, a dramatization of his individual contribution to spaceflight, and a set of motivations and emotions as complex and important to the narrative as those of the astronauts.

The Right Stuff and Motherhood in *Challenger Park*

In the opening scene of *Challenger Park*, Lucy Kincheloe is driving along NASA Road One near the Johnson Space Center, having left work hurriedly to get to her seven-year-old son's school after he has a minor asthma attack.[31] The

y of the scene is intentional; Lucy could be any worried mother in a mini-
ly one of a generation of women who grew up expecting to be able to have
both a challenging career and a satisfying family life. At a toy store with her son later that day, Lucy encounters a set of astronaut action figures, including one modeled after Christa McAuliffe. The figure haunts her:

> Lucy thought about the glossy, grinning replica of the Teacher in Space sealed in its transparent display package. She remembered an article she had read in the newspaper shortly after the *Challenger* disaster. A reporter had visited the McAuliffe home and spoken to the shell-shocked husband and mentioned a reverberant detail that had seared itself into Lucy's mind: a refrigerator door, crowded with magnets holding up children's artwork and reminders of soccer practices and orthodontist appointments that this devoted but now vanished mother would never keep. Christa McAuliffe's children had been nine and six. She had taken her son's stuffed frog into space.[32]

Some readers might wonder why, in a dual-career family in the twenty-first century, Lucy would interpret the McAuliffe children's soccer practices and orthodontist appointments as having been strictly maternal concerns. But Lucy considers her own children to be entirely her own domain; even as her mission draws near, we see her taking sole charge of domestic responsibilities such as hiring a new nanny, signing the children up for activities, and coordinating with her son's doctor over the treatment for his asthma. She never thinks of leaving those tasks to her spouse, as astronauts of earlier generations did. Lucy is apologetic and guilt-ridden when she has to depend on her husband to take care of simple chores like putting the children to bed while she is in the most intensive period of training for her mission; Brian's own long absences while training for and serving on his two missions (one of them a five-month stint on the International Space Station) are never described by Lucy, or anyone else, as impositions or as examples of his own failure as a parent. In fact, the child's asthma attack that opens the book occurs while Brian is on orbit, and Lucy seems unfazed to be handling the emergency on her own. Yet as her own mission approaches, her worst fear is being in space, unable to help, while her son has a medical crisis. (The fact that the child suffers from asthma gives poignancy and specificity to Lucy's anxiety; asthma provides a vivid metaphor for the airlessness of space, and Davis's futile gasping for breath during his asthma attacks echoes eerily one of the ways Lucy, who is training for a spacewalk, might die.)

The existence of women astronauts is both mundane and remarkable in *Challenger Park*. The first American women astronauts appeared when Lucy and her colleagues were children and young adolescents, and her lifelt dream of flying in space has always seemed within reach. Most Americans, including Lucy, attribute this change to general social and cultural pressures for inclusiveness, but viewed from another perspective, the inclusion of women in the astronaut corps, especially nonmilitary mission specialists, has as much to do with perceptions of the relative safety and simplicity of the spacecraft itself.

In a congressional hearing on the feasibility of including women as astronauts in 1962, Robert Voas, a training officer for Project Mercury, testified that the women should be patient and wait for the day when spaceflight had become safer and more routine:

> I think we all look forward to the time when women will be a part of our space flight team for when this time arrives, it will mean that man will really have found a home in space—for the woman is the personification of the home.[33]

Although shuttle commanders may argue that landing the space shuttle is more challenging than anything the Mercury astronauts were called upon to do, the fact that women can operate the shuttle changes the nature of the feat. As Weitekamp points out, "the very presence of women in orbit would indicate that space no longer remained a battlefield for international prestige."[34] That day seems to have arrived, but not without cost to the women who are finally allowed to participate in spaceflight. Lucy's overwhelming motivation in the novel is guilt—at every turn, she feels that her ambition to fly in space and the hours of training her mission necessitates are to the detriment of her children. While seeing her son into his school building, a space where her role is defined in terms of her motherhood, she recoils at her own official NASA portrait on the wall: "When she walked past the 'Astronaut Moms and Dads' photos, she glanced briefly at her own face, almost repelled by the bright, untroubled smile she saw there."[35] Here, and throughout the book, Lucy suspects she has no right to the public persona of astronaut in addition to her identity as a mother, and the narrative will prove her correct.

Lucy's constant guilt about shortchanging her children is only increased by the start of her affair with her training team supervisor, although her motivations for doing so are presented sympathetically. Lucy's husband, Brian, has shown himself to be a less-than-ideal astronaut on both his missions: during his stint on the International Space Station, he reacted to problems with an emotional rant on an open channel, and on a shuttle flight he misunderstood an order from Mission Control and caused a software problem that almost endangered the mission.[36] Although NASA, and Lucy, were willing to overlook his problems on one flight, two botched missions constitute proof positive that the Right Stuff. His chances of continuing a successful career as an [astron]ade, and as Lucy pulls away from him, she justifies her affair in terms [of his] failures. As in *The Right Stuff*, Brian's role as an astronaut is directly [tied to] his role as a husband—they are in fact one and the same. His failure in one role represents his unworthiness in both.

For Lucy, however, her role as an astronaut is not equated with her role as a wife—the two are constantly and visibly at odds. Without an earthbound spouse to take care of the domestic sphere and manufacture the aura of the Right Stuff for her, Lucy is stuck playing the roles of both warrior and worrier, a combination that is doomed to fail. The problems she faces in trying to do both recalls John Glenn's testimony at the congressional hearing on women astronauts in 1962. He argued to maintain a male-only group:

> I think this gets back to the way our social order is organized really. It is just a fact. The men go off and fight the wars and fly the airplanes and come back and help design and build and test them. The fact that women are not in this field is a fact of our social order. It may be undesirable.[37]

Margaret Weitekamp points out that Glenn's assertion of essentialism ("it is just a fact") is at odds with his attempt to justify and explain these roles—not to mention his willful ignorance of the Women Airforce Service Pilots and other women pilots who did, in fact, "fly the airplanes."[38] For Lucy, attempting to be one of the people who "goes off" leaves a gap to be filled in the home, a gap that Brian cannot properly fill. (This theme of women astronauts being unable to devote themselves to spaceflight because of their domestic duties was echoed in the media reaction to Lisa Nowak. A former NASA flight surgeon was quoted attributing Nowak's breakdown in part to the pressures on women astronauts: "They make more sacrifices than the 'Right Stuff' guys. They have to balance two careers: to be a mom and a wife and an astronaut. You don't come home at night, like most of the male astronauts, and have everything ready for you.")[39] Aside from guilt, Lucy's main emotion in the novel is fear: fear for her son's health, fear of leaving her children motherless, fear for her husband when he is traveling in space, then fear for herself. Of course, the role of astronaut's wife as designated worrier has been well established, and Lucy plays it perfectly when she gets word that her husband's mission has encountered a problem. Worried sick about him, she declares that she will not sleep until she learns that he is safe, but rather than waiting at home for the phone to ring, she drives to the space center and waits for hours in order to find out first-hand what is going on.[40] In her vigil, she resembles the Mercury and Apollo wives, those designated worriers of the previous generation, but her role is complicated by the fact that as an astronaut herself, she has access to restricted spaces and the knowledge to understand the technical talk swirling around her.

When it comes to her own mission, Lucy's fear becomes a looming emotion that threatens to swallow her entirely. In *Challenger Park*, astronauts' fear is more visible than in *The Right Stuff*, and after *Challenger*, the risks of spaceflight are more vivid and specific. Here Lucy's trainer and lover, Walt, describes why it is important to him to be on hand in Florida for the walkouts of the crews he has trained:

> The men and women sitting at the top of the stack, riding this pinnacle of fire as it burned through the sky, were courting perils that, even after *Challenger*, had still to be encountered and catalogued. More than one astronaut, Walt knew, had practiced his walk-out grin and his wave in front of the mirror, desperate not to let the blank terror show.[41]

In the space shuttle era, the astronauts work to hide their terror, but their terror does exist; for those with the Right Stuff, that fear would have been a sign of weakness, even if it were kept secret. That emotion was entirely in the domain of their wives, but in the shuttle era, the two roles have been collapsed into one.

In one of the scenes that sets the stage for their adulterous relationship, Lucy confides in Walt her fears about going to space, that she wants it to be over.

> "I want it to happen, then I want it to be behind me. I want to see my kids' faces when it's over. I want to hold them. I want to come back from the Cape and take a long bath and think about what it was like to have flown in space. Should I be telling you this? I don't have doubts, I really don't. It's just that every once in a while . . . every once in a while . . ."
>
> She leaned forward and whispered in a voice too low for [. . .] anyone else to hear.
>
> "I'm scared out of my mind."[42]

Lucy later regrets having told someone in a supervisory position about her fear, although Walt assures her that her feelings are normal, that astronauts who claim not to have felt any fear are not being entirely honest. But for Lucy this fear of the dangers of spaceflight blends with her maternal guilt to create a potent combination from which Mercury era astronauts were mercifully free. Lucy is hesitant to tell her children that she has been assigned to a mission, and when she does tell her seven-year-old son, he breaks down crying at the thought of his mother going into space.

> All at once she saw her destination through his innocent eyes, and pictured a realm that was familiar and alien at the same time,

infinitely black, infinitely still, the well of nothingness out of which he suspected he had risen into being, the same void that patiently waited to reclaim him and everyone he loved. Why would his mother, why would any mother, voluntarily leave her child to travel to such a place, a place that was as blank as death, and in whose perfect soundlessness his cries to her were sure to go unheard?[43]

Why would any mother, indeed? Yet fathers have been going to space since the very beginning, including this very child's father. Clearly, this astronaut's relationship with space can hardly be more different from those we see in Wolfe. Space is not only a symbol of death in the post-*Challenger* world, it is equated directly with death here in a simile. Lucy's desire to go to space in spite of the risks is not imagined as the conquering of a new frontier or as a victory of single combat or as a form of proof of her possessing anything like the Right Stuff. Rather, it is a failure of maternal duty, a selfish death wish. If the Right Stuff constituted a supreme type of courage, a willingness to risk one's life in order to benefit others, Lucy's urge toward spaceflight is the direct opposite—a desire to risk her life in spite of the effects on others.

The only other female astronaut we learn much about, Lucy's crewmate Patti, is single and childless, unencumbered by Lucy's guilt. Patti is described as being conventionally attractive, but her childlessness and her focus on her career make her unappetizing:

> Patti's eyes were tiny black staring buttons, and her constant powerful focus was indistinguishable from a trance. This distracting intensity, Lucy imagined, might have something to do with the fact that she was unmarried and unattached. No man, or woman for that matter, had been known to penetrate her force field. That sort of asexual bearing was certainly not something Lucy aspired to, but she could well understand its usefulness for an astronaut. As far as Lucy could tell, when the launch day came Patti would be leaving nothing behind. She would enter space without fear, possibly even without excitement, merely with the same attitude of blank contentment with which she experienced life on earth.[44]

Patti's competence, focus, and fearlessness make her sound quite a bit like the Mercury astronauts portrayed in *The Right Stuff*, competing to see whose heart rate was lowest during ignition and reentry. But in Patti, the Right Stuff is not only less than admirable, it is downright ugly. Unmitigated by motherhood and self-doubt, Patti's ambition makes her seem less than entirely human, and although Lucy is tortured by her own guilt and fear, Patti's form of the Right Stuff is nothing to envy.

Shortly before her launch, Lucy gives an interview to a female reporter who asks Lucy how she "find[s] her way as a woman in a culture that's so historically male. I mean, all those acronyms, all this vestigial macho *terseness*."[45] Lucy inwardly objects to the assumptions embedded in the question; as a well-trained astronaut, she knows the value of terseness, that her life might depend on it in an emergency. Lucy reflects that the reporter, being ten or twelve years older than her, must have been part of the feminist revolution and thus does not understand the complexities Lucy lives with.

> [Lucy] had entered the program at a time in the history of spaceflight when the idea of women astronauts was already so firmly established as to be mandatory. There were problems, still, of course, systematic male biases and presumptions common to most professions, but she did not care to advertise them to a stranger. Nor was she eager to let her interviewer know that most of the barricades she faced seemed pointedly of her own making.[46]

Lucy's career, as described in flashbacks throughout the book, has been exemplary. The only barricades of her own making, then, are her marriage and children. If Lucy is miserable, torn between her dream of flying in space and her desire to be a perfect mother to her children, her misery is understood to be her own fault. But the option to become a female astronaut and *not* have children is presented as an equally unnatural choice, as personified in the defeminized, soulless automaton Patti. (Marie Lathers points out a similar dichotomy between female astronaut as abandoning mother and female astronaut as unnatural woman in the media narratives around the deaths of Christa McAuliffe and Judith Resnik on *Challenger*.)[47] The only option for a woman, it seems, is to not have been an astronaut at all and to have stayed in the domestic sphere. John Glenn's statement that "it is just a fact of our social order" might sound outdated in the shuttle era, but Lucy's experience seems to justify it.

Geeks and Individuality in *Challenger Park*

Lucy's training team leader, Walt, is the sort of person who used to be passed over in spaceflight narratives. As we saw in The Right Stuff, "to be in engineering was to be an also-ran."[48] Norman Mailer, in *Of a Fire on the Moon*, a book of creative nonfiction on Apollo 11, writes

> The men who worked off NASA Highway 1 at the Manned Spacecraft Center were all clear-eyed and bullet-eyed and berry-eyed (pupils no larger than hard small acidic little berries) and they all seemed to wear dark pants, short-sleeve button-down white shirts and somber narrow ties.[49]

The description of bright and deadened eyes here sounds very similar to Harrigan's description of Patti's eyes in the voice of Lucy. In both cases, people who contribute to spaceflight in ways that do not match the simple Right Stuff model are scrutinized and found unnatural, as being less than fully human. Like Wolfe, Mailer is made uncomfortable by the engineers' interchangeability, by their value of collective effort over individual glory, and even by their pale skin; he notes that they "have a lunar air."[50]

But in the shuttle era, nonastronaut space workers can become human and even individual. One of the two point-of-view characters in *Challenger Park* is Walt, a midlevel worker at Johnson Space Center, one of tens of thousands of people working for NASA in jobs less glamorous than astronaut. The descriptions of Walt's life are intentionally bland: Walt lives in a nondescript condo in a nondescript suburb, has no hobbies and few friends, and finds satisfaction only in his work at Johnson Space Center. The only scenes in which Walt shows happiness are the moments when his team stays up all night working a problem, then goes out for breakfast. This narrative does not try to construct a heroic identity for Walt; his work is shared by a team, his life involves little risk, and he feels little desire to assert his masculinity. Yet in the shuttle era, this combination is more enticing to Lucy than is her astronaut husband.

In a quiet moment, Lucy gazes at a mural painted on a wall at the Johnson Space Center; and describing the image to herself allows her to reflect on the role of Walt and others like him (Figures 1 and 2):

FIGURE 1. Detail from Robert McCall's 1979 mural at Johnson Space Center as described by astronaut Lucy Kincheloe in Stephen Harrigan's novel *Challenger Park*: "a single woman astronaut positioned to the side of the central figures, her helmet in her hands and her hair in an antiquated flip." Courtesy artist Robert McCall; http://www.McCallStudios.com.

Among the people gazing heroically up to the heavens was a single woman astronaut positioned to the side of the central figures, her helmet in her hands and her hair in an antiquated flip. What kept Lucy from bristling at the image of a woman as a quaint variation in a robust male fantasy were the earnest tertiary figures stationed at the very bottom of the frame: flight controllers and trainers in their suits and white shirts and glasses . . . She was amused and touched, but maybe a little angry too, that in the bombastic scale of this painting he and his colleagues were forlorn marginal players. She thought Walt, with his Dockers and his pen-and-pencil set, should be the colossal figure in the middle of the canvas, in the same way he seemed to be slowly becoming the stable center of her own life.[51]

Lucy is happy to find men like Walt included in the painting, but feels annoyed that they are not given even higher billing, as high as the astronauts. In her own private life, Lucy is deceiving her husband, an astronaut, in order to be with Walt, whom she finds to be stronger, more stable, and more emotionally available. The definitions of masculinity have undergone a profound shift here if the Dockers-wearing NASA worker is more sexually desirable than the fighter-pilot-turned-astronaut to whom Lucy is married. In the shuttle era, technology is understood to be not just a means of getting single-combat warriors into battle but also the battle itself, and the men who master it are at the top of the new hierarchy.

FIGURE 2. The central section of Robert McCall's 1979 mural at Johnson Space Center showing "earnest tertiary figures stationed at the very bottom of the frame: flight controllers and trainers in their suits and white shirts and glasses." Courtesy artist Robert McCall; http://www.McCallStudios.com.

After an accident on her ill-fated flight, Lucy is stranded on the International Space Station for months before another shuttle can be sent to retrieve her. The accident Lucy encounters is surprisingly mundane; a meteorite strike causes fuel to vent onto Lucy's extravehicular activity (EVA) suit, and the shuttle must return to Earth before too much fuel is lost, leaving Lucy to let the fuel evaporate off her EVA suit before being let into the space station. In *The New York Times Book Review*, Thomas Mallon points out that "Harrigan keeps the thrills and derring-do to a sensible level. He knows that his real psychological business is to keep Lucy wan, anxious and depleted of bone mass—while back in Houston Walt holds a vigil whose true nature he must disguise from almost everyone."[52] It is interesting to note that even in her life-threatening emergency, Lucy is not portrayed as a hero, only as a physical object in the wrong place at the wrong time. In her moment of danger, the main drama does not come from Lucy's actions.

Although she begins her journey intending to leave her husband and start a new relationship with Walt, by the time she returns to Earth the crises she has faced on orbit and the time she is given to reflect have changed her mind. The advice her husband gives her based on his own time on the International Space Station is surprisingly helpful, and he shows himself to be a good father during a medical crisis while Lucy is in space. By the time she returns to Earth, Lucy has readjusted her values: after longing for home and missing her children for so many months, she will no longer make the mistake of prioritizing anything other than home and family.

When she explains her decision to Walt, she puts it not in terms of their relationship but in terms of the destruction her attempt to be an astronaut has already caused: "It was not in her, not really, to shatter her children's lives. It was not in Walt to demand such a thing. She had flown in space. She had held her children hostage to that dream, but she could not do it again for the sake of mere earthly happiness with a man who was not their father."[53] Her worst fear about spaceflight has come true: it did endanger her life and leave her children (temporarily) motherless. This was not the result when Brian flew on the shuttle, but it seems clear that mothers do not belong in space. This message is not lost on Lucy, who announces she does not want to fly on another mission. She will work to improve her marriage with Brian and keep her family together, and in fact, Brian seems revitalized by her return to Earth—he is no longer emasculated by Lucy's forcing him to take the worrier role. The apparent premise of the novel, that in the shuttle era an astronaut's wife can be an astronaut herself, has been negated. Lucy has been grounded, and she was wrong to have tried to fly at all.

Conclusion

Literature of spaceflight from the sixties era reflects a simple relationship between hero and public, a fascination with the figure of the astronaut as an avatar through whom readers could imagine the physical sensations and spiritual wonder of leaving the Earth. The astronauts with the Right Stuff were men with attributes we could never hope to emulate, and their wives were tasked with supporting their heroic performances, not only for the benefit of the astronauts themselves but also for the benefit of the public who needed to see the risk and daring performed through their worry. These early narratives are of risk and danger but never of disaster. The astronauts' roles as single-combat warriors, as military heroes, and as husbands and fathers situated them as symbolic figureheads of a specific narrative of American spaceflight: one of competitiveness, domination, and masculinity.

In the space shuttle era, astronauts' relationships with technology and spacecraft have changed, and readers' relationships with astronaut characters change as well. We can no longer indulge in simple hero-worship; astronauts are more like us, and spaceflight narratives now reflect a more complex emotional and cultural division of labor. An astronaut's spouse may be a man or may be a fellow astronaut, and so each astronaut is responsible for the manufacture and maintenance of his or her own Right Stuff. Both male and female astronauts are denied the simple heroism of the sixties era, and only partly because spaceflight is portrayed in the popular imagination as being more safe and routine (*Challenger* and *Columbia* notwithstanding). The only astronaut in *Challenger Park* who seems comfortable in his role, who seems to enjoy success and respect from his peers, is Lucy's commander, Surly Bonds—even his name evokes an old-school masculine (and military) representation of spaceflight. Surly is a former "top gun fighter pilot who had seen dozens of his friends auger into the ground but who nonetheless appeared to have coasted through life without an anxious thought."[54] This is textbook Right Stuff, and not coincidentally, Surly also has a traditional (nonastronaut) adoring wife who arranges life on Earth for him while he flies in space. In the shuttle era, women can be astronauts, but they cannot be entirely happy or entirely successful. They can remain single and childless, masquerading as men, or they can attempt to have it all, thereby emasculating their husbands and neglecting their children. Either way, they can never be heroes.

Rather than pointing out how heroic our heroes were, how amazing our amazingness, fiction in the space shuttle era, at least as represented in *Challenger Park*, turns these simple reflections upside down by making the extraordinary seem pedestrian while also making the everyday seem extraordinary.

Now that the space shuttle era is over, of course, the fiction that engages with American human spaceflight will by necessity become either increasingly historical or increasingly speculative, at least until the next stage of American spaceflight is well established enough for the next generation of novelists to take it on.

Notes

1. Jeff Prucher, *Brave New Worlds: The Oxford Dictionary of Science Fiction* (Oxford: Oxford University Press, 2007).

2. Alessandro Manzoni, *On the Historical Novel*, trans. Sandra Bermann (Lincoln: University of Nebraska Press, 1996), 23.

3. Tom Wolfe, *The Right Stuff* (New York: Farrar, Strauss, and Giroux, 1979), Kindle e-book; Norman Mailer, *Of a Fire on the Moon* (Boston: Little, Brown and Company, 1970); James Michener, *Space* (New York: Fawcett, 1983).

4. *Apollo 13*, DVD, directed by Ron Howard (1995; Universal City, CA: Universal Studios, 2006); *From the Earth to the Moon*, DVD, directed by David Frankel and Michael Grossman (1998; New York: HBO Home Video, 2009).

5. Stephen Harrigan, *Challenger Park* (New York: Knopf, 2006), Kindle e-book.

6. Other works of fiction from the space shuttle era (no prominent work of creative nonfiction has been published) include Douglas Coupland's *All Families Are Psychotic* (New York: Bloomsbury, 2001), Ira Sher's *Gentlemen of Space* (New York: Free Press, 2004), Patrick Ryan's *Send Me* (New York: Dial Press, 2006), and my own *The Time It Takes to Fall* (New York: Simon and Schuster, 2007).

7. Wolfe, *The Right Stuff*, 1.

8. Wolfe, *The Right Stuff*, 25.

9. Brian Abel Ragen suggests that the "malaise" and lack of common purpose of the 1970s are reflected as nostalgia in *The Right Stuff*, that "the account of the astronauts is by implication an attack on the culture of the 70s, a time when nothing would so engage everyone that traffic would stop all over the country, as it did when Alan Shepard was launched into space. *The Right Stuff* is a critique of the era in which it was written and a lament for an era that is irredeemably past." *Tom Wolfe: A Critical Companion* (Westport, CT: Greenwood Press, 2002), 138.

10. Wolfe, *The Right Stuff*, 16.

11. Wolfe, *The Right Stuff*, 352.

12. Ragen, *Tom Wolfe*, 122.

13. Wolfe, *The Right Stuff*, 57.

14. See Matthew H. Hersch's chapter in this volume.

15. Wolfe, *The Right Stuff*, 152.

16. Margaret Weitekamp, *Right Stuff, Wrong Sex: America's First Women in Space Program* (Baltimore: The Johns Hopkins University Press, 2004), 3.

17. Quoted in Marie Lathers, *Space Oddities: Women and Outer Space in Popular Film and Culture, 1960–2000* (New York: Continuum, 2010), 125.

18. Wolfe, *The Right Stuff*, 73.

19. Wolfe, *The Right Stuff*, 83.

20. Weitekamp, *Right Stuff, Wrong Sex*, 42.

21. Wolfe, *The Right Stuff*, 1.

22. Wolfe, *The Right Stuff*, 1.

23. Wolfe, *The Right Stuff*, 92.

24. Wolfe, *The Right Stuff*, 203.

25. Wolfe, *The Right Stuff*, 8.

26. Wolfe, *The Right Stuff*, 141.

27. Wolfe, *The Right Stuff*, 112.

28. See Charles Murray and Catherine Bly Cox, *Apollo: The Race to the Moon* (New York: Simon and Schuster, 1989).

29. Roger D. Launius, "Heroes in a Vacuum: the Apollo Astronaut as Cultural Icon" (paper presented at 43rd AIAA Aerospace Sciences Meeting and Exhibit, Reno, Nevada, January 10–13, 2005), 9.

30. Films of the space shuttle era seem to bear out *Challenger Park*'s ambiguous relationship with women astronauts: for instance, in the popular film *SpaceCamp* (directed by Harry Winer, 1986) a female shuttle pilot, Andie Bergstrom, is accidentally launched along with five teenagers, one of whom must take control and land the shuttle when Andie is injured.

31. Harrigan, *Challenger Park*, 3.

32. Harrigan, *Challenger Park*, 13.

33. Quoted in Weitekamp *Right Stuff, Wrong Sex*, 159. Also see discussion of women astronauts in Howard E. McCurdy, *Space and the American Imagination* (Washington, DC: Smithsonian Institution Press, 1997).

34. Weitekamp, *Right Stuff, Wrong Sex*, 159.

35. Harrigan, *Challenger Park*, 14.

36. Harrigan, *Challenger Park*, 34.

37. Quoted in Weitekamp, *Right Stuff, Wrong Sex*, 151.

38. Weitekamp, *Right Stuff, Wrong Sex*, 151.

39. Quoted in Amy E. Foster, *Integrating Women into the Astronaut Corps: Politics and Logistics at NASA, 1972–2004* (Baltimore: The Johns Hopkins University Press, 2011).

40. Harrigan, *Challenger Park*, 34.

41. Harrigan, *Challenger Park*, 18.

42. Harrigan, *Challenger Park*, 160–61.

43. Harrigan, *Challenger Park*, 110.

44. Harrigan, *Challenger Park*, 138–39.

45. Harrigan, *Challenger Park*, 188.

46. Harrigan, *Challenger Park*, 189.

47. Lathers, *Space Oddities*, 106–7.

48. Wolfe, *The Right Stuff*, 112.

49. Mailer, *Of a Fire on the Moon*, 11.

50. Mailer, *Of a Fire on the Moon*, 175.

51. Harrigan, *Challenger Park*, 187–88.

52. Thomas Mallon, "Satellite of Love," review of *Challenger Park*, by Stephen Harrigan, *The New York Times Book Review*, April 9, 2006, F7.

53. Harrigan, *Challenger Park*, 393.

54. Harrigan, *Challenger Park*, 112.

ACKNOWLEDGMENTS

I would like to particularly acknowledge Stephen J. Garber of NASA History Division. Steve participated in the early editing of the volume but because of the overwhelming press of work was forced to withdraw his name. His suggestions to authors on the first round of reviews were quite helpful. I would also like to thank Bill Barry, NASA Chief Historian, who supported my concept for the volume coming out of the "1961/1981: Key Moments in Human Spaceflight" conference and has provided further assistance since. Finally, I would like to thank the anonymous reviewers of the Smithsonian Institution Scholarly Press for their counsel.

Michael J. Neufeld
Editor

SELECTED BIBLIOGRAPHY

Allen, Graham. *Roland Barthes*. London: Routledge, 2003.

Andrews, James T. "Storming the Stratosphere: Space Exploration, Soviet Culture, and the Arts from Lenin to Khrushchev's Times." *Russian History* 36 (2009): 77–87.

———. *Red Cosmos: K. E. Tsiolkovskii, Grandfather of Soviet Rocketry*. College Station: Texas A&M University Press, 2009.

Andrews, James T., and Asif A. Siddiqi, eds. *Into the Cosmos: Space Exploration and Soviet Culture*. Pittsburgh, PA: University of Pittsburgh Press, 2011.

Atkinson, Joseph D., and Jay M. Shafritz. *The Real Stuff: A History of NASA's Astronaut Recruitment Program*. New York: Praeger, 1985.

Aubin, Jessica, Michelle Haak, and Andrew Mangini. "Media Coverage of Women Candidates." *White House Studies* 5, no. 4 (October 2005): 523–37.

Barghoorn, Frederick C. *Soviet Foreign Propaganda*. Princeton, NJ: Princeton University Press, 1964.

Barthes, Roland. *Mythologies*. 1957. Reprint, Paris: Seuil, 1970.

Becker, Christine. *It's the Pictures That Got Small: Hollywood Film Stars on 1950s Television*. Middletown, CT: Wesleyan University Press, 2008.

Bilstein, Roger E. *Flight in America: From the Wrights to the Astronauts*. Baltimore: Johns Hopkins University Press, 1984.

———. *Orders of Magnitude: A History of the NACA and NASA, 1915–1990*. Washington, DC: NASA, 1989.

———. *Stages to Saturn: A Technological History of the Apollo/Saturn Launch Vehicles*. Washington, DC: NASA, 1980.

Blackwood, Roy E. "The Content of News Photos: Roles Portrayed by Men and Women." *Journalism Quarterly* 60, no. 4 (Winter 1983): 710–14.

Boffey, Philip M. "NASA to Seek Observers to Fly on Shuttles." *New York Times*, December 16, 1983.

Brezhnev, Leonid I. *Leninskim kursom*. Volume 2. Moscow: Politizdat, 1974.

Broad, William J. "Reusable Space 'Truck' for Orbit Experiments." *New York Times*, April 7, 1984.

Burgess, Colin. *Selecting the Mercury Seven: The Search for America's First Astronauts*. Springer-Praxis Books in Space Exploration. New York: Springer, 2011.

Canby, Vincent. "'Big Screen' Takes on New Meaning." *New York Times*, April 19, 1987.
Carlier, Claude, and Marcel Gill. *Les trente premières années du CNES*. Paris: CNES/La documentation française, 1994.
Carter, Cynthia, and Linda Steiner. "Texts in Context." In *Critical Readings: Media and Gender*, edited by Cynthia Carter and Linda Steiner, 37–40. Maidenhead, UK: Open University Press, 2004.
Cartier, Raymond. "Le monde de demain commence aujourd'hui." *Paris Match*, no. 334, August 20, 1955, 54–59.
Chabbert, Bernard. *Les fils d'Ariane*. Paris: Plon, 1986.
Chang, Justin. Review of *Hubble 3D*. *Variety*, March 14, 2010.
Chrétien, Jean Loup. *Sonate au clair de terre: itinéraire d'un Français dans l'espace*. Paris: Denoël, 1993.
Clair, Benoît, and Patrick Baudry. *Aujourd'hui le soleil se lève 16 fois*. Paris: Carrère, 1985.
Collins, Martin. "Afterword: Community and Explanation in Space History (?)." In *Critical Issues in the History of Spaceflight*, edited by Stephen J. Dick and Roger Launius, 603–13. Washington, DC: NASA, 2006.
Collins, Michael. *Carrying the Fire: An Astronaut's Journeys*. New York: Farrar, 1974.
Cooper, Henry. *Before Lift-Off: The Making of a Space Shuttle Crew*. Baltimore: Johns Hopkins University Press, 1987.
Cooper, Nicola. "Heroes and Martyrs: The Changing Mythical Status of the French Army during the Indochinese War." In *France at War in the Twentieth Century: Propaganda, Myth, Metaphor*, edited by Valerie Holman and Debra Kelly, 126–41. New York: Berghahn, 2000.
Coupland, Douglas. *All Families Are Psychotic*. New York: Bloomsbury, 2001.
Cull, Nicholas J. *The Cold War and the United States Information Agency: American Propaganda and Public Diplomacy, 1945–1989*. New York: Cambridge University Press, 2008.
Cunningham, Walter. *The All-American Boys*. rev. ed. New York: ibooks, 2003.
Davis, Junetta. "Sexist Bias in Eight Newspapers." *Journalism Quarterly* 59, no. 3 (Autumn 1982): 456–60.
Dean, Margaret Lazarus. *The Time It Takes to Fall*. New York: Simon and Schuster, 2007.
Dean, Robert D. "Masculinity as Ideology: John F. Kennedy and the Domestic Politics of Foreign Policy." *Diplomatic History* 22 (Winter 1998): 29–62.
———. *Imperial Brotherhood: Gender and the Making of Cold War Foreign Policy*. Amherst: University of Massachusetts Press, 2001.

de Monchaux, Nicholas. *Spacesuit: Fashioning Apollo*. Cambridge, MA: MIT Press, 2011.

Dick, Steven J., ed. *Remembering the Space Age*. Washington, DC: NASA History Division, 2008.

Dick, Steven J., and Roger D. Launius, eds. *Critical Issues in the History of Spaceflight*. Washington, D.C.: NASA, 2006.

———, eds. *Societal Impact of Spaceflight*. Washington, DC: NASA History Division, 2007.

Dine, Philip. *French Rugby Football: A Cultural History*. New York: Berg, 2011.

Douglas, Susan J. *Where the Girls Are: Growing Up Female with the Mass Media*. New York: Times Books, 1994.

Douglas, Susan J., and Meredith W. Michaels. *The Mommy Myth: The Idealization of Motherhood and How It Has Undermined Women*. New York: Free Press, 2004.

Duncan, Margaret Carlisle. "The Politics of Women's Body Images and Practices: Foucault, the Panopticon, and Shape Magazine." *Journal of Sport and Social Issues* 18 (February 1994): 48–65.

Ebert, Roger. "*Hubble 3D*: A Journey into Time and Space." *Chicago Sun Times*, April 21, 2010.

Entrikin, J. Nicholas. *The Betweenness of Place: Towards a Geography of Modernity*. Baltimore: Johns Hopkins University Press, 1991.

Ezell, Edward C., and Linda N. Ezell. *The Partnership: A History of the Apollo-Soyuz Test Project*. Washington, DC: NASA History Division, 1978.

Faludi, Susan. *Backlash: The Undeclared War Against American Women*. New York: Anchor Books, 1992.

Ferguson, Marjorie. *Forever Feminine: Women's Magazine and the Cult of Femininity*. London: Heinemann, 1983.

Foote, John H. *Clint Eastwood: Evolution of a Filmmaker*. Westport, CT: Praeger, 2009.

Foster, Amy E. *Integrating Women into the Astronaut Corps*. Baltimore: Johns Hopkins University Press, 2011.

Gaffney, John, and Diana Holmes, eds. *Stardom in Postwar France*. Providence, RI: Berghahn, 2007.

Gagarin, Yuri. *Doroga v kosmose*. Moscow: Foreign Languages Publishing House, 1961.

Gauntlett, David. *Media, Gender and Identity: An Introduction*. London: Routledge, 2002.

Geppert, Alexander C. T., ed. *Imagining Outer Space: European Astroculture in the Twentieth Century*. New York: Palgrave Macmillan, 2012.

Gerovitch, Slava. "'New Soviet Man' Inside Machine: Human Engineering, Spacecraft Design, and the Construction of Communism," *Osiris*, 22 (2007): 135–57.

———. "The Human Side of a Propaganda Machine: The Public Image and Professional Identity of Soviet Cosmonauts." In *Into the Cosmos: Space Exploration and Soviet Culture*, edited by James T. Andrews and Asif A. Siddiqi, 77–106. Pittsburgh, PA: University of Pittsburgh Press, 2011.

Giddens, Anthony. *The Consequences of Modernity*. Stanford, CA: Stanford University Press, 1991.

Gill, Rosalind. *Gender and the Media*. Malden, MA: Polity Press, 2007.

Grosser, George H., Henry Wechsler, and Milton Greenblatt, eds. *The Threat of Impending Disaster, Contributions to the Psychology of Stress*. Cambridge: MIT Press, 1964.

Guibert, Louis. *Le vocabulaire de l'astronautique*. Rouen, France: Presses universitaires de Rouen, 1967.

Hansen, James R. *First Man: The Life of Neil Armstrong*. New York: Simon and Schuster, 2005.

Harp, Dustin, Jaime Loke, and Ingrid Bachmann. "First Impressions of Sarah Palin: Pit Bulls, Politics, Gender Performance, and a Discursive Media (Re)contextualization." *Communication, Culture, and Critique* 3 (2010): 291–309.

Harrigan, Stephen. *Challenger Park*. New York: Knopf, 2006.

Hecht, Gabrielle. *The Radiance of France, Nuclear Power and National Identity after World War II*. Cambridge, MA: MIT Press, 1998.

Heflick, Nathan A., and Jamie L. Goldenberg. "Sarah Palin, a Nation Object(ifie)s: The Role of Appearance Focus in the 2008 U.S. Presidential Election." *Sex Roles* 65 (2011): 149–55.

Heldman, Caroline, Susan J. Carroll, and Stephanie Olson, "'She Brought Only a Skirt': Print Media Coverage of Elizabeth Dole's Bid for the Republican Presidential Nomination." *Political Communication* 22 (2005): 315–35.

Hersch, Matthew H. "Spacework: Labor And Culture in America's Astronaut Corps, 1959–1979." PhD diss., University of Pennsylvania, 2010.

———. *Inventing the American Astronaut*. New York: Palgrave Macmillan, 2012.

Hixson, Walter L. *Parting the Curtain: Propaganda, Culture, and the Cold War, 1945–1961*. New York: St. Martin's Press, 1997.

Hodnett, Grey, ed. *Resolutions and Decisions of the Communist Party of the Soviet Union*. Vol. 4, *The Khrushchev Years, 1953–1964*. Toronto: University of Toronto Press, 1974.

Horrigan, B. "Popular Culture and Visions of the Future in Space, 1901–2001." In *New Perspectives on Technology and American Culture*, edited by B. Sinclair, 49–67. Philadelphia: American Philosophical Society, 1986.

Hughes, Thomas P. *American Genesis: A Century of Invention and Technological Enthusiasm, 1870–1970.* New York: Penguin Books, 1990.

Inness, Sherrie A. *Tough Girls: Women Warriors and Wonder Women in Popular Culture.* Philadelphia: University of Pennsylvania Press, 1999.

Jenks, Andrew. "Yuri Gagarin and the Search for a Higher Truth." In *Into the Cosmos: Space Exploration and Soviet Culture,* edited by James T. Andrews and Asif A. Siddiqi, 107–32. Pittsburgh, PA: University of Pittsburgh Press, 2011.

———. *The Cosmonaut Who Couldn't Stop Smiling: The Life and Legend of Yuri Gagarin.* DeKalb: Northern Illinois University Press, 2012.

Jones, Ray, Audrey J. Murrell, and Jennifer Jackson. "Pretty Versus Powerful in the Sports Pages: Print Media Coverage of U.S. Women's Olympic Gold Medal Winning Teams." *Journal of Sport and Social Issues* 23 (May 1999): 183–92.

Josephson, Paul R. *Would Trotksy Wear a Bluetooth: Technological Utopianism under Socialism, 1917–1989.* Baltimore: Johns Hopkins University Press, 2010.

Kelly, Catriona, and Mark Bassin, eds. *Soviet and Post-Soviet Identities.* Cambridge: Cambridge University Press, 2012.

Kevles, Daniel J. *The Physicists: The History of a Scientific Community in Modern America.* Cambridge: Harvard University Press, 1995.

Khrushchev, Nikita S. "On Peaceful Coexistence." *Foreign Affairs,* 38, no. 1 (October 1959): 1–18.

Kilgore, De Witt Douglas. *Astrofuturism: Science, Race, and Visions of Utopia in Space.* Philadelphia: University of Pennsylvania Press, 2003.

Kuisel, Richard. *Seducing the French: The Dilemma of Americanization.* Los Angeles: University of California Press, 1993.

Kuhn, Annette, ed. *Alien Zone: Cultural Theory and Contemporary Science Fiction Cinema.* New York: Verso, 1990.

Larres, Klaus, and Kenneth A. Osgood, eds. *The Cold War after Stalin's Death: A Missed Opportunity for Peace?* Lanham, MD: Rowman and Littlefield, 2006.

Lathers, Marie. *Space Oddities: Women and Outer Space in Popular Film and Culture, 1960–2000.* New York: Continuum, 2010.

Launius, Roger D. "A Western Mormon in Washington, D.C.: James C. Fletcher, NASA, and the Final Frontier." *Pacific Historical Review* 64 (1995): 217–41.

———. "Heroes in a Vacuum: The Apollo Astronaut as Cultural Icon." *Florida Historical Quarterly* 87 (Fall 2008): 174–209.

———. *After Apollo: The Legacy of the American Moon Landings.* New York: Oxford University Press, 2012.

Launius, Roger D., John M. Logsdon and Robert W. Smith, eds. *Reconsidering Sputnik: Forty Years Since the Soviet Satellite.* Harwood Academic, 2000.

le Bailly, Jacques. "Canaveral français au Sahara." *Paris Match*, no. 519, March 21, 1959, 64–69.

Lefebvre, Henri. *The Production of Space*. Translated by Donald Nicholson-Smith. London: Blackwell, 1991.

Lenin, Vladimir I. *Collected Works*. Moscow: Progress Publishers, 1964.

Lerner, Warren. "The Historical Origins of the Soviet Doctrine of Peaceful Coexistence." *Law and Contemporary Problems* 29, no. 4 (Autumn 1964): 865–70.

Lewis, Cathleen S. "From the Cradle to the Grave: Cosmonaut Nostalgia in Soviet and Post-Soviet Film." In *Remembering the Space Age*, edited by Steven J. Dick, 253–70. Washington, DC: NASA, 2008.

Logsdon, John M. *John F. Kennedy and the Race to the Moon*. New York: Palgrave Macmillan, 2010.

Logsdon, John M., Roger D. Launius, David H. Onkst, and Stephen J. Garber, eds. *Exploring the Unknown: Selected Documents in the History of the U.S. Civil Space Program*. Washington, DC: NASA, 1995.

Lucanio, Patrick, and Gary Colville. *American Space Science Fiction Television Series of the 1950s: Episode Guides and Casts and Credits for Twenty Shows*. Jefferson, NC: McFarland and Company, 1998.

Mailer, Norman. *Of a Fire on the Moon*. Boston: Little, Brown and Company, 1970.

Mallon, Thomas. "Satellite of Love," review of *Challenger Park*, by Stephen Harrigan. *The New York Times Book Review*, April 9, 2006, F7.

Manzoni, Alessandro. *On the Historical Novel*. Translated by Sandra Bermann. Lincoln: University of Nebraska Press, 1996.

Markula, Pirkko. "Firm but Shapely, Fit but Sexy, Strong but Thin: The Postmodern Aerobicizing Female Bodies." *Sociology of Sport Journal* 12 (1995): 424–453.

Markwick, Roger D. "Peaceful Coexistence, Detente and Third World Struggles: The Soviet View, from Lenin to Brezhnev." *Australian Journal of International Affairs* 44, no. 2 (1990): 171–94.

Marx, Karl, and Friedrich Engels. *The Communist Manifesto*. London: Penguin Classics, 2002.

Maurer, Eva, Julia Richers, Monica Rüthers, and Carmen Scheide. *Soviet Space Culture: Cosmic Enthusiasm in Socialist Societies*. London: Palgrave Macmillan, 2011.

Matheson, Richard. *Richard Matheson's The Twilight Zone Scripts: Volume Two*. Edited by Stanley Wiater. Springfield, PA: Edge Books, 2002.

Mathias, Paul, and Marc Heimer. "Gemini X: Un triomphe, mais pourquoi plus un russe dans l'espace depuis seize mois?" *Paris Match*, no. 903, July 30, 1966: 16–18.

May, Elaine Tyler. *Homeward Bound: American Families in the Cold War Era.* New York: Basic Books, 1988.

McCray, Patrick. *Keep Watching the Skies! The Story of Operation Moonwatch and the Dawn of the Space Age.* Princeton, NJ: Princeton University Press, 2008.

McCurdy, Howard E. *Space and the American Imagination.* Washington, DC: Smithsonian Institution Press, 1997.

McDougall, Walter A. "Space-Age Europe: Gaullism, Euro-Gaullism and the American Dilemma." *Technology and Culture* 26 (1985): 179–203.

———. *The Heavens and the Earth: A Political History of the Space Age.* Baltimore: Johns Hopkins University Press, 1997.

McLuhan, Marshall. *Understanding Media: The Extensions of Man.* New York: McGraw-Hill, 1964.

Michael, D. N. "The Beginning of the Space Age and American Public Opinion." *Public Opinion Quarterly* 24 (1960): 573–82.

Michener, James. *Space.* New York: Fawcett, 1983.

Mindell, David A. *Digital Apollo: Human and Machine in Spaceflight.* Cambridge: MIT Press, 2008.

Mumford, Lewis. "No: 'A Symbolic Act of War...'" *New York Times*, July 21, 1969.

Murray, Charles, and Catherine Bly Cox. *Apollo: The Race to the Moon.* New York: Simon and Schuster, 1989.

Nelson, Ronald Roy, and Peter Schweizer. *The Soviet Concepts of Peace, Peaceful Coexistence, and Détente.* Lanham, MD: University Press of America, 1988.

Neufeld, Michael J. *Von Braun: Dreamer of Space, Engineer of War.* New York: Knopf, 2007.

O'Leary, Brian. "Topics: Science or Stunts on the Moon?" *New York Times*, April 25, 1970.

Ordway, F. I., III, and R. Lieberman, eds. *Blueprint for Space: Science Fiction to Science Fact.* Washington, DC: Smithsonian Institution Press, 1992.

Osgood, Kenneth A. *Total Cold War: Eisenhower's Secret Propaganda Battle at Home and Abroad.* Lawrence: University of Kansas, 2006.

Oushakine, Serguei Alex. *The Patriotism of Despair: Nation, War, and Loss in Russia.* Ithaca, NY: Cornell University Press, 2009.

Perry, Geoffrey E. "Perestroika and Glasnost in the Soviet Space Programme." *Space Policy* 5, no. 4 (November 1989): 279–87.

Philmus, Lois C. *A Funny Thing Happened on the Way to the Moon.* New York: Spartan Books, 1966.

Pop, Virgiliu. "Viewpoint: Space and Religion in Russia: Cosmonaut Worship to Orthodox Revival." *Astropolitics* (May 2009): 150–63.

Portree, David S. F. *NASA's Origins and the Dawn of the Space Age.* Monographs in Aerospace History 10. Washington, DC: NASA, 1998.

Potter, W. James. "Gender Representation in Elite Newspapers." *Journalism Quarterly* 62, no. 3 (Autumn 1985): 636–40.

Prucher, Jeff. *Brave New Worlds: The Oxford Dictionary of Science Fiction*. Oxford: Oxford University Press, 2007.

Ragen, Brian Abel. *Tom Wolfe: A Critical Companion*. Westport, CT: Greenwood Press, 2002.

Rey, H. A. *Curious George Gets a Medal*. Boston: Houghton Mifflin, 1957.

Richmond, Yale. *Cultural Exchange and the Cold War: Raising the Iron Curtain*. University Park: Pennsylvania State University Press, 2003.

———. *Practicing Public Diplomacy: A Cold War Odyssey*. New York: Berghahn Books, 2008.

Rieffel, Rémy."Les charactéristiques et la spécificité de la presse magazine en France." In *Die Zeitschrift—Medium der Moderne*, edited by Clemens Zimmermann and Manfred Schmeling, 43–62. Frankreich-Forum 6. Saarbrücken: Universität des Saarlandes, 2005.

Rockwell, Trevor. "The Molding of the Rising Generation: Soviet Propaganda and the Hero-Myth of Iurii Gagarin," *Past Imperfect*, 12 (2006): 1–34.

Roland, Alex. "Barnstorming in Space: The Rise and Fall of the Romantic Era of Spaceflight, 1957–1986." In *Space Policy Reconsidered*, edited by Radford Byerly Jr., 33–52. Boulder, CO: Westview Press, 1989.

Rosen, Ruth. *The World Split Open: How the Women's Movement Changed America*. New York: Penguin Books, 2000.

Ross-Nazzal, Jennifer. "Détente on Earth and in Space: The Apollo-Soyuz Test Project," *OAH Magazine of History* 24, no. 3 (1 July 2010): 29–34.

Rouse, Morleen Getz. "A History of the F. W. Ziv Radio and Television Syndication Companies: 1930–1960." PhD diss., University of Michigan, 1976.

Rozwadowski, Helen M. "Small World: Forging a Scientific Maritime Culture for Oceanography." *Isis* 87 (1996): 409–29.

Ryan, Patrick. *Send Me*. New York: Dial Press, 2006.

Sagdeev, Roald, and Susan Eisenhower. "United States-Soviet Space Cooperation during the Cold War." NASA. http://www.nasa.gov/50th/50th_magazine/coldWarCoOp.html (accessed January 15, 2010).

Santy, Patricia A. *Choosing the Right Stuff: The Psychological Selection of Astronauts and Cosmonauts*. Westport, CT: Praeger, 1994.

Schwartz, Harry. "Space Program: Behind the Triumph, Criticism of Goals." *New York Times*, August 17, 1969.

Shayler, David J. *Skylab: America's Space Station*. Chichester, UK: Praxis, 2001.

———. *NASA's Scientist-Astronauts*. New York: Springer, 2007.

Sheehan, Michael J. *The International Politics of Space*. New York: Routledge, 2007.

Sher, Ira. *Gentlemen of Space*. New York: Free Press, 2004.

Siddiqi, Asif A. *Challenge to Apollo: The Soviet Union and the Space Race, 1945–1974*. Washington, DC: NASA History Division, 2000.

———. *Sputnik and the Soviet Space Challenge*. Gainesville: University Press of Florida, 2003.

———. *The Soviet Space Race with Apollo*. Gainesville: University Press of Florida, 2003.

———. *The Red Rocket's Glare: Spaceflight and the Soviet Imagination, 1857–1957*. Cambridge: Cambridge University Press, 2010.

———. "Cosmic Contradictions: Popular Enthusiasm and Secrecy in the Soviet Space Program." In *Into the Cosmos: Space Exploration and Soviet Culture*, edited by James T. Andrews and Asif A. Siddiqi, 47–76. Pittsburgh, PA: University of Pittsburgh Press, 2011.

Simon, Rita J., and Jean M. Landis. "The Polls—A Report Women's and Men's Attitudes about a Woman's Place and Role." *Public Opinion Quarterly* 53 (1989): 265–76.

Skuridin, G. A., et al. *Entrance of Mankind into Space (15th Anniversary of the First Manned Flight into Space)*. NASA Technical Translation NASA-TT-F-17114. Washington, DC: NASA, 1976.

Slayton, Donald K., and Michael Cassutt. *Deke! U.S. Manned Space: From Mercury to the Shuttle*. New York: St. Martin's Press, 1994.

Smith, Charles W. "It's Colossal! It's Stupendous! It's IMAX!" *Reader's Digest*, August 1985, 87.

Smith, Michael. "Selling the Moon: The U.S. Manned Space Program and the Triumph of Commodity Scientism." In *The Culture of Consumption: Critical Essays in American History, 1880–1980*, edited by Richard Wightman Fox and T. J. Jackson Lears. New York: Pantheon Books, 1983.

Staff of the *Washington Post*. *Challengers: The Inspiring Life Stories of the Seven Brave Astronauts of Shuttle Mission 51-L*. New York: Pocket Books, 1986.

Sterling, Christopher H., and John Michael Kittross. *Stay Tuned: A History of American Broadcasting*. 3rd ed. Mahwah, NJ: Lawrence Erlbaum Associates, 2002.

Stites, Richard. *Revolutionary Dreams: Utopian Vision and Experimental Life in the Russian Revolution*. Oxford: Oxford University Press, 1989.

Stuhlinger, Ernst, and Frederick I. Ordway III. *Wernher Von Braun: Crusader for Space*. Malabar, FL: Krieger, 1996.

Suid, Lawrence H. *Guts and Glory: The Making of the American Military Image in Film*. Lexington: University Press of Kentucky, 2002.

Swenson, Loyd S., James M. Grimwood, and Charles C. Alexander. *This New Ocean: A History of Project Mercury*. Washington, DC: NASA, 1966.

Tuan, Yi-Fu. *Space and Place: The Perspective of Experience*. Minneapolis: University of Minnesota Press, 1977.

U.S. Department of State. *Foreign Relations of the United States, 1958–1960*. Vol. 2, *United Nations and General International Matters*. Washington, DC: U.S. Government Printing Office, 1991.

———. *Foreign Relations of the United States, 1961–1963*. Vol. 5, *Soviet Union*. Washington, DC: U.S. Government Printing Office, 1998.

———. *Foreign Relations of the United States, 1961–1963*. Vol. 25, *Organization of Foreign Policy; Information Policy; United Nations; Scientific Matters*. Washington, DC: U.S. Government Printing Office, 2001.

———. *Foreign Relations of the United States, 1969–1976*. Vol. 1, *Foundations of Foreign Policy, 1969–1972*. Washington, DC: U.S. Government Printing Office, 2003.

von Braun, Wernher. *The Mars Project*. Urbana: University of Illinois, 1953.

von Braun, Wernher, and Cornelius Ryan. *Conquest of the Moon*. New York: Viking Press, 1953.

Weinberg, Alvin M. "Impact of Large-Scale Science on the United States." *Science* 134 (1961): 161–64.

Weitekamp, Margaret A. *Right Stuff, Wrong Sex: America's First Women in Space Program*. Baltimore: Johns Hopkins University Press, 2004.

Whipp, Glenn. Review of *Hubble 3D*. *Los Angeles Times*, March 19, 2010.

Wolf, Naomi. *The Beauty Myth: How Images of Beauty Are Used Against Women*. New York: William Morrow, 1991.

Wolfe, Tom. *The Right Stuff*. New York: Farrar, Strauss, and Giroux, 1979. Kindle e-book.

Zubok, Vladislav. *Zhivago's Children: The Last Russian Intelligentsia*. Cambridge, MA: Harvard University Press, 2009.

CONTRIBUTORS

Margaret Lazarus Dean holds a BA in anthropology from Wellesley College and an MFA in creative writing from the University of Michigan. She is the author of *The Time It Takes to Fall*, a novel about the space shuttle *Challenger* disaster. A book of creative nonfiction, *Leaving Orbit: Notes from the Last Days of American Spaceflight*, will be published by Graywolf Press in 2014. She is an assistant professor of English at the University of Tennessee and lives in Knoxville.

Guillaume de Syon teaches history at Albright College in Reading, Pennsylvania, and is a history research associate at Franklin & Marshall College in Lancaster, Pennsylvania. He is the author of *Zeppelin! Germany and the Airship 1900–1939* and of the textbook *Science and Technology in Modern European History*. His current research focuses on the role of media and advertising in the public understanding of technology.

Matthew H. Hersch is a lecturer in science, technology, and society in the Department of History and Sociology of Science at the University of Pennsylvania, where he received his PhD. During his doctoral studies, he held the 2009–10 HSS-NASA Fellowship in the History of Space Science and a 2007–8 Guggenheim Fellowship at the Smithsonian Institution's National Air and Space Museum. During the production of this book, he served as the Postdoctoral Teaching Fellow for the Aerospace History Project of the Huntington-USC Institute on California and the West. He is the author of *Inventing the American Astronaut*.

Andrew Jenks is an associate professor of history at California State University, Long Beach. He is the author of numerous articles on Russian history, the history of technology, and environmental history as well as the cofounder of the Russian History Blog (http://russianhistoryblog.org/). He is also the author of *Russia in a Box: Art and Identity in an Age of Revolution*, *Perils of Progress: Environmental Disasters in the Twentieth Century*, and *The Cosmonaut Who Couldn't Stop Smiling: The Life and Legend of Yuri Gagarin*.

Valerie Neal is a curator in the Division of Space History at the Smithsonian's National Air and Space Museum. Since 1989, her responsibilities have included the Skylab, space shuttle, and Spacelab artifact collections, and she has led the development of three major long-term exhibitions on human spaceflight. She writes and publishes on space shuttle topics and was instrumental in the refurbishment of shuttle orbiter *Enterprise* and the acquisition of shuttle orbiter *Discovery*. She earned a PhD in American Studies at the University of Minnesota.

Michael J. Neufeld is a museum curator in the Space History Division of the National Air and Space Museum, Smithsonian Institution. From 2007 to 2011 he served as division chair. Born and raised in Canada, he has four history degrees, including a PhD from Johns Hopkins University in 1984. He has written three books, *The Skilled Metalworkers of Nuremberg: Craft and Class in the Industrial Revolution*, *The Rocket and the Reich: Peenemünde and the Coming of the Ballistic Missile Era*, and *Von Braun: Dreamer of Space, Engineer of War*. He has edited three others, *Planet Dora* by Yves Béon, *The Bombing of Auschwitz: Should the Allies Have Attempted It?* (with Michael Berenbaum), and *Smithsonian National Air and Space Museum: An Autobiography* (with Alex Spencer).

Trevor S. Rockwell has published articles on the proliferation of science and technology themes in American propaganda and how closely Yuri Gagarin's autobiography complied with Soviet propaganda directives. His dissertation, "Space Propaganda 'For All Mankind': Soviet and American Responses to the Cold War, 1957–1977," compares space exploration narratives in American and Soviet propaganda magazines. He currently teaches American history and the history of science and technology at the University of Alberta, Edmonton, Canada, where he received his PhD.

Jennifer Ross-Nazzal is the historian for the NASA Johnson Space Center. She completed her baccalaureate degree at the University of Arizona and master's in history at New Mexico State University. She holds a PhD in history from Washington State University. She has authored and coauthored numerous articles about the space shuttle, the NASA community, and the Johnson Space Center. She is also an accomplished oral historian. She is the author of *Winning the West for Women: The Life of Suffragist Emma Smith DeVoe*.

James Spiller is an associate professor of history at the College at Brockport, State University of New York. He teaches courses about the modern United States, as well as the histories of American environment, culture, economics, and science and technology. His nearly completed manuscript *Frontiers for the*

American Century examines the cultural politics of U.S. space and Antarctic exploration throughout the cold war. He is also Brockport's incoming dean of the Graduate School.

Margaret A. Weitekamp is a curator in the Division of Space History at the Smithsonian's National Air and Space Museum, where she oversees over 4,000 individual pieces of space memorabilia and space science fiction objects, including toys and games, clothing and stamps, medals and awards, and buttons and pins, as well as comics and trading cards. She is the author of *Right Stuff, Wrong Sex: America's First Women in Space Program*, which won the Eugene M. Emme Astronautical Literature Award given by the American Astronautical Society. She earned her BA at the University of Pittsburgh and her PhD at Cornell University.

INDEX

Numbers in *italic* text indicate pages with illustrations

Abbey, George, 156
Aelita, 84
African Americans
  diversity of population and ability to relate to astronauts, 66
  space shuttle astronaut recruitment and opportunities, 67, 71, 162, 190
  support roles at NASA, 66, 67
Agnew, Spiro, 68
Air Force, U.S.
  experimental aircraft development and testing, 38–39, *39*
  movies and television programs, involvement in production of, 15, 35, 53n19
  X-15 program and full-pressure flight suit development, 19
Aldrin, Edwin "Buzz," Jr, 137
Alien, 46
Allen, Joseph P. "Joe," 54n35, 72–73, 152, 172n21, 196
Allen, Woody, 42
American Museum of Natural History, 151
Amerika Illiustrirovannoye (*America Illustrated*) magazine
  cooperative space programs, 134–35, 145n57
  cosmonaut visits to US, coverage of, 138, 147n87
  danger narratives, 128–29
  deaths of spacefarers, 128–29
  failures and setbacks, publication of information about, 128, 142n19
  families and home life of spacefarers, representation of, 129–30, 138, *139*
  features and format of, 125–26
  openness of space program to the media, 127, 128, 134, 137–38, 141nn12–13
  peaceful and warm space explorers, representation as, 126, 129–30
  peaceful coexistence propaganda message of, 5, 135–40, 147nn85–87
  publication, distribution, and sales of, 125–26, 140n4
  spacefarers, representation of, 5, 126–31
  transformation of humanity through spaceflight, 131–35, 139–40, 143–44nn42–43
  writers for, 140n1
Ananoff, Alexandre, 109–10, 112
Apollo 1 mission, 119
Apollo 8 mission, 67, 118
Apollo 9 mission, 118
Apollo 10 mission, 118
Apollo 11 mission
  astronauts from, appearance before congress by, 127
  funding for space programs, reinstatement of, 68
  live broadcasts of moonwalk, 129
  Mobile Quarantine Facility use following, 128
  Paris Match coverage about, 118
  space race and, 2
Apollo 12 mission, 118–19
Apollo 13 (film), 49, 50, 51, 76, 170, 204, 213
Apollo 13 mission, 65, 119
Apollo program
  Apollo-Soyuz Test Project, 2, 120, 134–35, 138–39
  dangers and risks of spaceflight, 173n48
  fictional characterizations of, 204
  "Men of the Year" designation, 67
  movies, depiction of in, 48
  Paris Match coverage about, 118–19
  scientist-astronaut mission, 44–45
  spacecraft design, 2
Armstrong, Neil
  biography of, 7n9
  giant leap quote, 140, 144n43
  Islamic call to prayer, hearing on the moon, 7n9
  post-Apollo 11 image of, 7n9
  test pilot role of, 35–36, *37*
Army, U.S., 38
Asian Americans, astronaut opportunities for, 71, 162
astroculture, 6

astronauts
    announcement of first, 29
    celebrity status and media coverage of, 1–4, 57–58, 114–18, 123n39, 126–31
    characteristics, image, and appearance of, 1–4, 36–37, *37*, 128
    daily life of, 138, 150, 163
    deaths of, 2, 119, 128–29, 142n22
    families and home life of spacefarers, representation of, 63, 129–30, 138, *139*, 143n38, 209–11
    female astronauts, prohibition against, 66
    fictional characterizations of, 3–4, 6, 38, 40–42, 204
    first astronauts, selection of, 39–40
    frontier and pioneer images of, 4, 52, 58–67, 76
    hero status of, disappearance of, 5, 121
    identification of in spacesuits, 19–20
    interest in and research on by historians and cultural scholars, 1, 3–4, 7n7
    loneliness experienced by, 129
    loner and self-reliant qualities, 65
    memoirs of, 4
    mission assignments, 43
    peaceful and warm space explorers, representation as, 5, 126
    peaceful coexistence propaganda message and representation of, 137–38, 147nn85–87
    photographs of and public appearances by, 127–28, 141n13
    pilot-hero images of, 1–4, 5, 6, 18, 36–37, 40–43, 46, 50, 161–64, 205–8
    public enthusiasm for, 57–58
    qualities and physical capabilities needed for spaceflight, 22–28, 29, 37, 46–52, 64, 71–73, 72, 76–77, 207–9
    religious beliefs of and expressions of faith by, 137–38
    selection of, 43–44, 51
    terminology and naming of, 40
    test pilot and fighter pilot background of, 1–2, 29, 35–36, 38–40, 51–52, 66
    See also scientist-astronauts; space shuttle astronauts; women astronauts
Aurora 7 spacecraft, 128, 130

Barthes, Roland, 109, 118
Barton, Glenn (George Nader), 20–24, *23*, 25, 27–28
Baudry, Patrick, 120, 124n62

Beggs, James, 47, 153
Bell Aircraft, *39*
Béon, Yves, 113
Beregovoi, Georgi, 138
Berkner, Lloyd, 60
Berry, Michael A., 182
Bewitched, 33n51
biographies of astronauts, 4, 7n9
Blackwood, Roy E., 178
Blaha, John, *160*
Blériot, Louis, 111
Blue Planet, 165, *168*, 169, 172n18, 173n28, 173n44
Bluford, Guion S. "Guy," 162, 179, 190
Bogdanov, Aleksander, 84
Bolden, Charlie, 173n44
Bonanza, 14
Bond, James, 48
Bonestell, Chesley, 11, 15, 110, 122n15
Borman, Frank, 118
Boyne, Walter, 154
Bradley, Truman, 14
Brischel, Francine, 192
Bronson, Charles, 42
Buchli, Jim, 173n44
Buck Rogers, 4, 12
Buck Rogers in the 25th Century, 48
Burroughs, Edgar Rice, 59
Burtt, Ben, 173n28
Bush, George H. W., 75–76
Bush, Vannevar, 60
Bykovsky, Valery, *131*

Canby, Vincent, 151
Captain Video and His Video Rangers, 4, 10–11, 12, 38
Carpenter, Robin, 130
Carpenter, Scott, 117, 128, 130
Carter, Jimmy, 70
Cartier, Raymond, 110, 111, 117–18
Cedar Point, 171n13
Centre National d'Etudes Spatiales (CNES), 109, 111, 121
Chaffee, Roger, 142n22
Challenger
    accident with and loss of crew of, 2, 47–48, 74–76, 164, 166, 196
    crew of, *75*
    The Dream is Alive screenings and accident with, 166
    IMAX filming during missions, 166, 172n22

Challenger Park (Harrigan)
    emotional and cultural issues, interpretation of through, 6, 204–5, 213, 219–24
    fear, expression of, 217
    geeks, engineers, and accomplishments of spaceflight, 213, 219–21, *221*
    individuality and single-combat warfare, 220–21
    motherhood and career balance, representation of, 213–19, 225n30
Chapman, Philip, 54n35
chimpanzees for first flights, 207–8
Chinese space program, 2
Chrétien, Jean-Loup, 120, 121
Christ Appears to the People (Ivanov), 97
Circus World, 171n13
The Cisco Kid, 14
Clair, Benoît, 120
Cleave, Mary L., 192
Coats, Michael L., 192
Cobb, Jerrie, 207
Collier's magazine, 11, 110, 122n15
Collins, Michael, 40, 152–53, 170
Columbia, 49, *49*, 57–58, 167
Commando Cody, Sky Marshal of the Universe, 12
Conquest of Space, 38
Conrad, Jane, 209
Conrad, Peter, 208, 209
Cooper, Leroy Gordon, Jr., 50, 128
cosmism and cosmist rapture, 93–98
Cosmonaut Pavilion, Exhibit of People's Economic Achievement, 81, *82*, 102, 102n2
cosmonauts
    celebrity status and media coverage of, 1–4, 114–18, *115*, *116*, 126–31
    characteristics, image, and appearance of, 1–4, 128
    deaths of, 2, 119, 128–29, 142n22
    families and home life of spacefarers, representation of, 129–30
    fighter pilot background of, 1
    hero status of, 137
    interest in and research on by historians and cultural scholars, 1, 3
    loneliness experienced by, 129
    media coverage of US visits by, 138, 147n87
    peaceful and warm space explorers, representation as, 5, 126
    peaceful coexistence propaganda message and representation of, 135–40, 146n71, 146n81
    socialism and symbols of peace, 126–27, 135–36, 140–41nn6–7, 146n71
    training of, 128, 141n18
Crippen, Robert, *49*, 161–62, 173n36, 190, 192, 195
Cronkite, Walter, 158, 163–64, 166
Crossfield, Scott, 19
Cruise, Tom, 166–67
Cuban invasion and the Bay of Pigs, 28
Cunningham, Walter, 44, 48, 54n37, 147n87
Curious George Gets a Medal, 43

David Clark flight suits, 19
Davis, Junetta, 178
Dean, Robert D., 25, 28
Deep Impact, 49–50
Defense, U.S. Department of, 9, 15, 20, 30
de Gaulle, Charles, 111, 114
Destination Moon, 11, 38
Destiny in Space, 165, 167–68, 169, 172n18, 173n45
DiCaprio, Leonardo, 164
Dickinson, Angie, 17
Discovery, 160, 172n22
Disney, Walt, 11, 59
Disneyland, 58–59, 138, 147n87
Disneyworld, 138, 147n87
Dobrovolski, Georgi, 142n22
Donner, Richard, 42, 53n26
Douglas, David, 173n28
The Dream is Alive
    astronaut as cinematographers, 155–61, *160*, 172–73nn27–28, 173n30, 173n36
    astronauts, representation in, 149–51, *150*, 162–65, 170, 173n44, 173n46
    Challenger accident and screening of, 166
    dangers and risks of spaceflight, acknowledgment of, 165–67
    identity and personalities of astronauts, 164–65, 170, 173n44, 173n46
    narration of, 158, 163–64, 166
    presentation and showings of, 149
    production of and filming footage for, 149–51, 154–61, 171n5, 171nn2–3, 172n18, 172nn21–22
    public enthusiasm for, 150, 151, 168–69
    records about, 171n3
    reviews of, 151, 169, 174n52
    theater attendance statistics, 168–69, 170–71n1

The Dream is Alive (continued)
   themes, images, and content on, 5, 71, 157–61
   title of, 172n20
DuMont Network, 10–11, 12
Dunbar, Bonnie J., 192
Duvall, Robert (Spurgeon Tanner), 49–50

"Earthrise" photo, 118
Eastwood, Clint, 50, 51
Ebert, Roger, 151–52
Ebony magazine, 66
Eisenhower, Dwight D., 39–40, 51, 110, 137
engineers
   geeks, engineers, and accomplishments of spaceflight, 211–13, 219–21, *221*
   heroic role for, 211–12
England, Anthony, 54n35
Engle, Joseph "Joe," 45, 49
Enlightenment, 126, 131, 135, 139
Enterprise, 47
European Space Agency, 73, 75
Expo '74, 152–53, 171n13

Fabian, John M., 188, 191
Faith 7 spacecraft, 128
Fantastic Voyage, 45
FBI, involvement in television program production, 16
Feoktistov, Konstantin, 138
Ferguson, Graeme, 152–54, 156, 159, 164, 171n3, 172nn18–19
Ferguson, Marjorie, 177
Ferguson, Phyllis, 171n3
Ferraro, Geraldine, 195
films (movies)
   combat films, 40–41, 53n19
   government department involvement in production of, 35, 53n19
   pilot-astronauts, representation of, 48–51
   piloting skills, representation of, 49–51
   scientists and scientist-astronauts, representation of, 43, 45–46
   spacecraft concepts and characteristics of crews, 37–39, *39*, 47–48, 52n5
   stereotypes introduced and reinforced in, 40–42, 48–49
   See also IMAX films
Fisher, Anna L.
   appearance, weight, and physical fitness of, *176*, 177, 179–82, 189–90, 198–99nn30–31

   leave of absence from flying, 196
   marriage of, 183–84
   mission assignment and media coverage of, 196
   motherhood, 186, *187*, 196
Fisher, Kristin, 186, *187*
Fisher, William F. "Bill," 183–84, 186, *187*
Flash Gordon, 12
Fletcher, James, 46, 69, 70
Fooner, Andrea, 181
Ford, Gerald, 70
Forever Feminine (Ferguson), 177
France
   agreements with to fly a Frenchman into orbit, 120
   cold war and foreign policy of, 120
   French identity and culture, manufacturing and delivery of, 108–9, 121
   heroes in, 114, 120–21
   nuclear program in, 109, 122n8
   space program in, 109, 111
Freeman, MacGillivray, 152
Friendship 7 spacecraft, 65, 128
From the Earth to the Moon, 204, 213
frontier and pioneer images, 4, 58–67, 76
Frosch, Robert A., 175
Fullerton, Gordon, 47

Gagarin (Gzhatsk), Russia, 86, 87–88, 89–92, 98–100, 101–2, 106n75
Gagarin, Yuri
   attitudes toward and memories about, 81–83, 100–102, 102nn2–3
   autobiography and biography of, 4, 7n9, 136
   background and early life of, 5, 82, 84, 86–89
   celebration of flight as triumph of socialism, 126–27, 135–36, 140–41nn6–7
   cosmism and cosmist rapture, 93–98
   date of first flight by, 1, 28
   death of, 97, 119, 128
   family of, *91*, 92–93, *93*, 101
   flight of and public expectations for a better life, 83–86
   length of flight, mystical significance of, 86
   Moon, plans for exploration of, 144n46
   movie about, 81, 102n3
   museums, tourism, and celebrations related to early life of, 5, 82, 87–96, *87*, *90*, *91*, *95*, 98–100, *99*, 101–2, 105n56, 106n75
   myth and national treasure status of, 4–5, 81–83, 86, 93–98, 100–102, 126–27

Paris Match coverage about, 115
photo of, *131*
poetry written about, 85, 97, 101
posters of and displays of images of, 85–86
reaction to flight by in US, 10
religious beliefs of, 96–98
seeing God in Space, 97
selection of for first flight, 4–5, 82
Gagarina, Elena, 101, 102n3
Gagarin's Grandson, 81, 102n3
Galileo spacecraft, 196–97
Gardner, Dale A., 196
Garn, Jake, 196
Garneau, Marc, 173n36, 195
Garner, James, 50
Garriott, Owen, 54n34
Gemini program
  fictional characterizations of, 204
  Paris Match coverage about, 117
  photos of astronauts and families, *139*
  spacecraft design, 2
gender roles
  breaking down of, 25
  diversity of astronaut trainees and crews, 67, 71, 76, 162–63, 175, 190
  flight malfunctions and the difference between male and female astronauts, 194
  qualities and physical capabilities needed for spaceflight, 22–28, 29
  The Right Stuff and representation of, 209–11
  television programs and depiction of, 29–30, 33n51
  See also masculinity and masculine ideals; women astronauts
Gerard, Gil, 48
Gibson, Edward, 54n34
Gibson, Paul, 185–86
Gibson, Robert L. "Hoot," 185, 186, *188*
Giddens, Anthony, 109, 121
Glenn, John, Jr.
  duty to inspire young men, 65
  fictionalization of shuttle flight of, 50
  Friendship 7 spacecraft, 65, 128
  handshake with Titov and cooperative space program, 145n57
  military pilot background of, 40
  women astronauts, opinion about, 216
"Going to Work in Space" program, 150, 171n4
Goodrich flight suits, 19
Gordon, Theodor, 112

Graham, Billy, 62, 63
Granville, Joan, 24
Graveline, Duane, 54n34
Grissom, Virgil "Gus," 36, 44, 128, 141n13, 142n22
Groupement Astronautique Français, 109–10
Guillaumat, Pierre, 111
Gunsmoke, 14
Gzhatsk (Gagarin), Russia, 86, 87–88, 89–92, 98–100, 101–2, 106n75

Hail Columbia! 153, 154, 161–62, 166, 172n18, 172n20
Haise, Fred, *47*
Hallyday, Johnny, 123n36
Ham (chimpanzee), flight of, 208
Hanks, Tom, 49, 213
Harrigan, Stephen, 6, 204–5
Hartsfield, Henry W., 162, 193, 194
Hauck, Frederick H., 192
Hawley, Steven A. "Steve," 163, 192
The Heavens and the Earth (MacDougall), 51
Heimer, Marc, 111, 112, 115, 119
Henize, Karl, 54n35
Herbert, Charles, 17
Hergé, 119
heroic era of human spaceflight, 2, 6n1, 204
Hersch, Matthew H., 4
Hines, Bill, 175
Hispanics, astronaut opportunities for, 162
history and historical fiction, difference between, 203–4
Holdren, Judd, 12
Holmquest, Donald, 54n35
Homeward Bound (May), 25
Hubble 3D, 151–52, 156, 164, 165, 167, 168, 169, 171n3, 172n18, 173n45
Hubble Space Telescope, 197
Hudson, Rock, 28
human spaceflight
  competitive space programs, 2
  conference to commemorate first flight, 1
  cooperative space programs, 2, 134–35, 145n57
  criticism of programs, 46, 55n54
  dangers and risks associated with, 2, 5, 9–10, 17, 18, 29, 40–42, 128–29, 142n22, 151, 165–67, 173n48
  fictional characterizations of, 3–4, 6, 38, 40–42, 204

human spaceflight (*continued*)
    geeks, engineers, and accomplishments of, 211–13, 219–21, *221*
    global culture of space, understanding of, 107, 121
    heroic era of, 2, 6n1, 204
    hero status of spacefarers, disappearance of, 5, 121
    public enthusiasm for, 3, 51
    rationales for and value of space exploration, 17–18, 29, 51, 131–35, 143–44nn42–43, 144n51
    research and science mission and objectives, 44, 46
    routine nature of, 121
    transformation of humanity through, 125–26, 131–35, 139–40, 140n1, 143–44nn42–43
    See also astronauts; cosmonauts

I Dream of Jeannie, 29, 33n51
I Led Three Lives, 15–16
IMAX Corporation, 149, 153, 171n3
IMAX films
    3-D films, production of, 168
    astronauts, representation in, 5, 149–51, *150*, 161–65, 170, 173n44, 173n46
    astronauts as cinematographers, 155–61, *160*, 172–73nn27–28, 173n30, 173n36, 173n45
    dangers and risks of spaceflight, acknowledgment of, 151, 165–67
    development of technology for, 151, 152, 171n3
    hair and hairstyles and IMAX cameras, 194
    origin of, 152–54
    positive and upbeat themes, 151, 154–55, 165–66
    production of and filming footage for, 5, 149–51, 152–61, 171n5, 172n18, 172nn21–22, 173nn44–46, 174nn49–50
    profits from films, 172n19
    public enthusiasm for, 149, 150, 151–52, 153, 168–69
    records about, 171n3
    reviews of, 151–52, 169, 174n52
    space shuttle program, promotion of, 71, 153–54
    theater attendance statistics, 168–69, 170–71n1
    theater system and technology for showing, 152–53, 171n11, 171n13

themes, images, and content on, 149–50, 157–61, 167–70, *168*, *169*
    See also *The Dream is Alive*
International Astronautical Congress, 109–10
International Astronautical Federation, 109–10
Interplanetary Revolution, 84
Inventing the American Astronaut (Hersch), 4
Ivanov, Alexander, 97
Ivins, Marsha, 157

Japanese space program, 73, 75
The Jetsons, 29
Johnson, Lyndon B., 29
Johnson Space Center mural, 220–21, *220*, *221*
Jones, Tommy Lee, 50, 51

Kennedy, Jackie, 114, 117
Kennedy, John F.
    foreign policy of and response to threats by, 25, 28
    popularity of in France, 114
    space policy and programs, 28–29, 40, 51, 137
Kerwin, Joseph, 54n34
Khrushchev, Nikita, 3, 97, 127, 136, 141n7
King of the Rocket Men, 12
Klushino, Russia, 82, 88–89, 91–92, *91*
Komarov, Vladimir, 119, 142n22
Korolev, Sergei, 130, 144n46
Kraft, Christopher C., 175, 197n3
Krikalev, Sergei, *169*
Kubrick, Stanley, 45

Larsen, Keith, 26
Lathers, Marie, 1, 3
Launius, Roger, 3, 144n51, 213
Lawrence, John, 189
Leestma, David C., 166, 195
Lenoir, William, 54n35
Leonov, Alexei, 101, 115–16, *115*, *116*, 117, 121, 147n87
Leskov, L. V., 94
Ley, Willy, 11, 38, 122n15
Liberty Bell 7 spacecraft, 128
Life magazine
    astronauts, promotion of, 40, 63, 64–65, 66, 107
    circulation and availability of, 119
    decline in sales of, 119
    Paris Match, syndicated stories for, 111–12, 114
Lindbergh, Charles, 65
Living History Center, 171n13

Llewellyn, John, 54n35
Lockheed Corporation/Lockheed Martin, 149, 153–54, 172nn18–19
Lovelace Clinic, 207–9
Lovell, James "Jim," 49, 51
Lucas, George, 45
Luce, Henry, 114
Lucid, Michael, 183
Lucid, Shannon
    appearance, weight, and physical fitness of, 176
    domesticity, homemaking skills, and motherhood, 182–83, 184–85, 199n43
    IMAX films, involvement in production of, 165
    media coverage of, 175, 177
    mission assignment and media coverage of, 196–97
    qualities and physical capabilities needed for spaceflight, 189, *189*
Lundigan, William "Bill" (Edward McCauley), 16–20, *19*, 24, 26–27, *26*, 41

MacDougall, Walter, 51
MacLeish, Archibald, 125, 126, 140n1
Mailer, Norman, 204, 219–20
The Making of an Ex-Astronaut (O'Leary), 42
The Man and the Challenge
    masculinity and masculine ideals, representation of, 22–24, 26, 27–28, 29
    production of, 14, 20
    reception to and early space television programming, 10
    sets and costumes, 20–21, 22
    stereotypes introduced and reinforced in, 4
    storyline and fact-based subject matter, 9–10, 20–25, *21*, 29–30
    women, representation of, 23–24
Manzoni, Alessandro, 203–4
maritime culture and scientists, 45, 54n46
Martin Company, *39*
masculinity and masculine ideals
    American space program success and, 28–29
    bare-chested display of, 22–23, *23*
    Mercury astronauts, 63–65, 205–8
    pilot-hero image of astronauts, 1–4, 5, 6, 18, 205–8
    popular culture and masculinity models, 25–28
    television programs and depiction of, 4, 22–28, 29

Matheson, Richard, 41
Mathias, Paul, 111
Mattingly, Thomas, 50
May, Elaine Tyler, 25
McAuliffe, Christa, 48, 73, 74, 214, 219
McCall, Robert, 137–38, 220
McCandless, Bruce, 172n21
McCauley, Edward (Bill Lundigan), 16–20, *19*, 24, 26–27, *26*, 41
McCormack, John, 127
McCurdy, Howard, 7n7, 11, 48–49, 121
McDivitt, James, *139*
McDivitt, Patricia, *139*
McLuhan, Marshall, 107
Meadows, Joyce, 23–24
media
    celebrity status of spacefarers, 1–4, 57–58, 114–18, *115*, *116*, 123n39, 126–31
    feminism, the women's movement, and attitudes toward women, 177–79
    openness of US space program to the media, 127, 128, 134, 137–38, 141nn12–13
    print media importance during age of few televisions and government control of channels, 107, 108, 121, 124n63
    women astronauts, reception of, 5–6, 130–31, 143n38, 175–79, 190–97, 197n2, 197n5, 200n76
    women's magazines, 177–78, 180–82, 198–99nn30–31
Medvedev, Dmitrii, 102
Men into Space
    masculinity and masculine ideals, representation of, 26–27, 29
    rationales for and value of space exploration, 17–18, 29
    reception to and early space television programming, 10
    sets and costumes, 15, *16*, 19–20, *19*
    stereotypes introduced and reinforced in, 4, 40–41
    storyline and fact-based subject matter, 9–10, 15–20, *16*, 29–30
    women, representation of, 23–25
Mercury astronauts
    celebrity status and media coverage of, 2, 3, 57–67
    diversity of population and ability to relate to, 66
    duty to inspire young men, 65–66

Mercury astronauts (*continued*)
  families and home life of spacefarers, representation of, 209–11
  fictional characterizations of, 204 (see also *The Right Stuff* [Wolfe])
  frontier and pioneer images of, 4, 58, 61–67, 76
  hero status and pilot-hero images of, 4, 65–66, 76
  masculinity and masculine ideals, 63–65, 205–8
  movies, depiction of in, 48, 50
  Paris Match coverage about, 117
  photos of, 62, *64*
  promotion of, 40
  qualities and physical capabilities needed for spaceflight, 63–65, 64, 71–72, 207–9
  selection of, 39–40
  spacesuit development for, 19
  training of, 63, 64, 72
Mercury program spacecraft, 2
Michel, Frank Curtis, 54n34
Michener, James, 204
Military Academy, U.S., 19, 31n18
Miller, Mark, 28
mission looking for science and science looking for mission, 46, 55n54
mission specialists, 5
Mission to Mars, 50–51
Mission to Mir, 165, 169, 172n18
Mona, Project (Pioneer), 112, 123n24
Moon
  Apollo 11 mission to land on, 2, 68
  Kennedy space policy and decision to land on, 28–29, 51
  lunar lander design, 2
  Mobile Quarantine Facility use following mission to, 128
  scientist-astronaut missions, 45
  Soviet missions to orbit and land on, 2
  Soviet plans for exploration of, 132, 144n46
  television programs and depiction of, *16*, 17–18
Moon (film), 52
Moonraker, 48
Moore, Mary Tyler, 42
"Moral Code of the Builder of Communism," 136
movies. *See* films (movies); IMAX films
Mueller, George, 68
Mullane, Richard M. "Mike," 163, 192
Musgrave, F. Story, 54n35

Myers, Toni, 156, 164, 165, 173n46
My Favorite Martian, 29

Nader, George (Glenn Barton), 20–24, *23*, 25, 27–28
National Aeronautics and Space Administration (NASA)/National Advisory Committee for Aeronautics (NACA)
  astronauts, announcement of first, 29
  astronauts, selection of first, 39–40
  budgets and funding for programs, 66–67, 68–69, 70
  experimental aircraft development and testing, 38–39, *39*
  films, concerns about accuracy of, 153–54
  frontier and pioneer images, 60–61
  "Going to Work in Space" program, 150, 171n4
  IMAX films, involvement in production of, 149, 153–54, 172n18
  mission and objectives of, 44, 46, 60–61, 68–69
  movies and television programs, involvement in production of, 35
  X-15 program and full-pressure flight suit development, 19
National Commission on Space, 74
National Geographic magazine, 63, 70
national heros, 57–58, 67–68, 76–77
A Nation at Risk, 73–74
Navy, U.S., 19, 38
Neihouse, James, 160, 173n28
Newell, Homer, 45
Niebuhr, Reinhold, 61–62
Nikolayev, Andrian, 130, *131*
Nikolayeva, Elena, 131
"1961/1981: Key Moments in Human Spaceflight" conference, 1
Nixon, Richard, 46, 65, 68–69, 70, 140n1, 143n42
North American Aviation, 35
Nowak, Lisa, 216

Of a Fire on the Moon (Mailer), 204, 219–20
O'Leary, Brian, 42, 54n35
O'Neill, Gerard, 69–70, 71, 74
Osgood, Kenneth, 138

Paine, Thomas, 68, 69, 74
Pal, George, 11
Paresev 1-A vehicle, *36*
Paris Match magazine
  advertising style of, 108

Apollo program, coverage of, 118–19
astronauts, syndicated stories about, 111–12, 114, 123n39
astronauts and cosmonauts, coverage about, 114–18, 123n39
Bonestell illustrations in, 110, 122n15
circulation and availability of, 108, 119, 121n3
deaths, coverage of, 119
decline in sales of, 5, 119
features and format of, 108–9, 120–21
French identity and culture, manufacturing and delivery of, 108–9, 121
French spationauts, coverage about, 5, 120–21, 124n62
front covers of, 110, 111, 114, 115–16, *115*, 118, 119, 120, 123n36
global culture of space, understanding of, 107, 121
human element, importance of and focus on, 110–12, 121
Life, syndicated stories from, 111–12, 114
middle class values, presentation of, 109
presentation of events in, 109, 117–18
spaceflight and space travelers coverage in, 5, 107, 108, 109–21, 122n5
space race, coverage about, 117–18
technology and machines, coverage about, 108, 109, 110
translation of US stories, 111–12
von Braun, coverage about and articles written by, 112–14, 123n33, 123nn35–36
writers for, 111
Parker, Robert, 54n35
Patsayev, Viktor, 142n22
Peck, Gregory, 41, 53n19
pilot-hero image of astronauts, 1–4, 5, 6, 18, 36–37, 40–43, 46, 50, 161–64, 205–8
Pioneer (Project Mona), 112, 123n24
Pioneering the Space Frontier, 74
pioneer settlers and frontier images, 58, 67–76
The Poem of Ecstasy (Scriabin), 83
Popovich, Pavel, *131*, 143n38
Popovicha, Marina, 143n38
Potter, W. James, 178
Protazanov, Iakov, 84
Prouvost, Jean, 111
Putin, Vladimir, 102

Quaid, Dennis, 50

radio programming and productions, 10, 13, 108
Ragen, Brian Abel, 206–7
Reagan, Ronald, 70–71, 74, 75, 196
Recer, Paul, 182–83
Redbook magazine, 180–82, 198–99nn30–31
Red Planet, 52
Red Star, 84
religious beliefs and expressions of faith, 96–98, 137–38, 146n81
Resnik, Judith "Judy"
    appearance, weight, and physical fitness of, *176*, 193–94, *193*, 200n76
    death of, 166, 196
    domesticity and homemaking skills, 182
    The Dream is Alive filming, 163, 166
    flight malfunctions and the difference between male and female astronauts, 194
    hair and hairstyles for astronauts, 194
    ladies, offensive reaction to use of term, 188
    marriage interests of, 184
    media coverage of, 177
    mission assignment and media coverage of, 192–94, 200n76
    privacy and personal life of, 193–94
Reuben H. Fleet Space Theater and Science Center, 171n13
Ride, Sally
    appearance, weight, and physical fitness of, *176*, 179
    Challenger accident investigation commission, 196
    dating and marriage, 184
    domesticity and homemaking skills, 182
    first American woman in space, 162
    media coverage of, 175, 177
    mission assignment and media coverage of, 190–92, *191*, 195, *195*
The Right Stuff (film), 48, 50, 162
The Right Stuff (Wolfe)
    character, personal behaviors, and images of astronauts, representation of, 3–4, 48, 205–9
    communism and collective work, anxiety about, 212–13
    creative nonfiction account of Mercury program, 205
    emotional and cultural issues, interpretation of through, 6, 204, 224n9
    families and home life of spacefarers, representation of, 209–11

The Right Stuff (Wolfe) *(continued)*
   geeks, engineers, and accomplishments of spaceflight, 211–13
   gender roles and division of labor in, 209–11
   individuality and single-combat warfare, 211–13
Right Stuff, Wrong Sex (Weitekamp), 207
The Road to the Stars (Gagarin), 136
Robertson, Cliff, 12
Rocketship X-M, 38
rocket sled experiments, 20–21, *21*
Rockwell, Sam, 52
Rocky Jones, Space Ranger, 38
Rod Brown of the Rocket Rangers, 12
Roddenberry, Gene, 31n18
Roosevelt, Theodore, 4, 58, 61–62, 65, 66, 67, 75
Rozwadowski, Helen, 45
Russia
   Cosmonaut Pavilion, Exhibit of People's Economic Achievement, 81, *82*, 102, 102n2
   discoveries by and accomplishments of ethnic Russians, 84, 95–96
   empowerment of through Gagarin flight and memory, 100–101
   folk traditions and culture, revival of, 86–87
   Russian space program and cooperative space programs, 2, 134–35, 145n57
   See also Soviet Union

Sagdeev, Roald, 134
Salyut 1 space station, 119
Salyut 6 space station, 120
Saratov, Russia, 81, *82*, 87, *90*, 94, *95*
Schmitt, Harrison "Jack," 44–45, 54n34
Schweickart, Russell, 91
science dividend, 118
science fiction and imaginary space travel
   depictions of space travelers in, 1, 3
   early space television programming depiction of spaceflight, 10–12
   frontier and pioneer images, 59
   interplanetary travel, depiction of, 59
   Russian stories and films, 84
   spacecraft concepts and characteristics of crews, 37–39, *39*, 48–49, 52n5
Science Fiction Theater, 14
science looking for mission and mission looking for science, 46, 55n54
Science: The Endless Frontier (Bush), 60

scientist-astronauts
   acceptance of and reception to by pilot-astronauts, 43–45, 54n37
   analysis of data from missions, 46, 55n54
   Apollo flight flown by, 44–45
   fictional spacecraft concepts and characteristics of crews, 37–39, *39*, 52n5
   marginalized or secondary roles of, 4, 18
   motivation of for flying in space, 46
   pilot-hero image of astronauts and, 2–3, 37
   popular culture, television, and representation of, 42–43, 45–46
   public enthusiasm for and acceptance of, 42–46
   qualities and physical capabilities needed for spaceflight, 37, 42–46
   selection of, 43–44, 54nn34–35
   Skylab missions, 54n41
Scobee, Dick, 166
Scott, Ridley, 45–46
Scriabin, Alexander, 83
Scripps Institute of Oceanography, 16
Scully-Power, Paul, 173n36
Sea Hunt, 16
Seddon, Rhea M.
   appearance, weight, and physical fitness of, *176*, 177, 179–80, 181, *188*
   dating and marriage, 184
   domesticity and homemaking skills, 182
   mission assignment and media coverage of, 196–97
   mission assignments for women, 192
   motherhood, 185–86
Shepard, Alan
   mission assignments, responsibility for, 43
   Paris Match coverage about, 117
   photographs of and public appearances by, 141n13
   reaction to flight by in US, 10, 65, 224n9
   success of flight and US space policy, 28–29
Shepard, Bill, *169*
Shepard, Louise, 117, 210–11
Shepard, Sam, 50
Sinise, Gary, 50
Skerritt, Tom, 50
Skylab missions
   daily life of astronauts, 138, 150
   Paris Match coverage about, 119–20
   scientist-astronaut missions, 54n41
Skylab program, 2

Slayton, Donald "Deke," 40, 43, 44, 45
Smith, Greg, 173n28
Smithsonian National Air and Space Museum, 149, 151, 152–54, 169, 171n13, 172n18
socialism and symbols of peace, 126–27, 135–36, 140–41nn6–7, 146n71
society and culture
  astroculture, 6
  celebrity status of astronauts in popular culture, 1–4, 57–58
  diversity of population and ability to relate to astronauts, 66, 76
  enthusiasm for space programs by general public, 3, 51
  frontier and pioneer images, 4, 58–67, 76
  masculinity models and popular culture, 25–28
  national heroes, 57–58, 67–68, 76–77
  pioneer settlers and frontier images, 58, 67–76
  propaganda and cultural representation of astronauts, 1–3
Soros, George, 105n56
Soviet Embassy, Washington, DC, 125
Soviet Life (*USSR*) magazine
  cooperative space programs, 134–35, 145n57
  danger narratives, 128–29, 141n18
  deaths of spacefarers, 128–29
  earlier name of, 140n3
  events to honor cosmonauts, coverage of, 126–27, 140–41nn6–7
  families and home life of spacefarers, representation of, 129–30
  features and format of, 125–26
  peaceful and warm space explorers, representation as, 126, 129–30
  peaceful coexistence propaganda message of, 5, 135–40, 146n71, 146n81
  publication, distribution, and sales of, 125–26, 140nn3–4
  spacefarers, representation of, 5, 126–31, 140–41nn6–7, 141n9
  transformation of humanity through spaceflight, 131–35, 139–40
  women, representation of, 130–31, *131*, 143n38
Soviet Union
  Afghanistan, invasion of, 120
  collapse of, 81
  collapse of and tourism surrounding Gagarin myth, 5, 82–83
  Cosmonaut Pavilion, Exhibit of People's Economic Achievement, 81, *82*, 102, 102n2
  cultural and social history, interest in, 3
  demise of, 2
  "Moral Code of the Builder of Communism," 136
  premonitions of space and expectations for a better life in, 83–86
  religious beliefs and expressions of faith in, 96–98, 137, 146n81
  technological utopianism, 83–84
  See also Russia
Soviet Union space program
  cooperative space programs, 2, 134–35, 145n57
  decline of programs, 118
  failures and setbacks, secrecy about, 141n18
  French agreement with to fly a Frenchman into orbit, 120
  Moon, plans for exploration of, 2, 132, 144n46
  public enthusiasm for, 3
  secrecy surrounding flights, 2, 3, 117–18, 119, 134, 141n18
  space stations missions, 2, 73, 119, 120, 132
  Sputnik, 15, 51, 60, 61
  superpower space race for firsts, 2, 117–18
  world leadership and success of, 60–61, 75
  See also cosmonauts
Soyuz 1 mission, 142n22
Soyuz 11 mission, 119, 142n22
Soyuz spacecraft
  Apollo-Soyuz Test Project, 2, 120, 134–35, 138–39
  design of, 2
Space (Michener), 204
Space and the American Imagination (McCurdy), 7n7, 11
SpaceCamp, 47, 50, 225n30
Space Cowboys, 49, 50, 51, 76
spacecraft
  airplane-like qualities of, 46–52
  Apollo program, 2
  capsule-ballistic missile concepts, 39
  Gemini program, 2
  Mercury program, 2
  science fiction and imaginary space travel, 37–39, *39*, 48–49, 52n5
  Soviet spacecraft, 2
  terminology for, 40

Space Exploration Initiative, 75–76
Spacelab Life Sciences 1 (SLS-1) mission, 197
Space Oddities (Lather), 1, 3
Space Patrol, 11, 12
Spaceship Earth, 125
space shuttle astronauts
   celebrity status and media coverage of, 57–58
   daily life of, 163
   diversity of trainees and crews, 67, 71, 76, 162–63, 175, 190
   fictional characterizations of, 223–24 (see also *Challenger Park* [Harrigan])
   flight malfunctions and the difference between male and female astronauts, 194
   French spationauts, 120, 124n62
   IMAX films, representation in, 5, 149–51, *150*, 161–65, 170, 173n44, 173n46
   pilot-astronauts, representation of in movies and television programs, 48–51
   pilot-hero image of, 5
   piloting skills, representation of in movies and television programs, 49–51
   pilots for flying shuttle, 47, *47*
   pioneer settlers and frontier images of, 4, 52, 58, 67–76
   public enthusiasm for, 75
   qualities and physical capabilities needed for spaceflight, 46–52, 71–73, *72*
space shuttle program
   accidents with shuttles, 2, 47–48, 74–76, 164, 166–67, 196
   announcement of, 46, 68–69
   authorization to build and start of flights with, 2
   budgets and funding for, 47–48, 66–67, 68–69, 70
   concept behind and missions of, 46–47
   conference to commemorate first flight, 1
   crew, nontraditional and nonprofessional, 46–47, 73
   dangers and risks of spaceflight, 165–67
   fictional characterizations of, 204–5
   "Going to Work in Space" program, 150, 171n4
   IMAX films to promote, 71, 153–54
   public enthusiasm for, 71
   reliability of shuttles, 73
Space Station 3D, 165, 166–67, 168, 169, *169*, 172n18, 174nn49–50

space stations, Soviet missions to, 2, 73, 119, 120, 132
Space: The New Frontier, 58–59
Spaulding, M. C., 15
Spock, Mr., 45
Sputnik, 15, 51, 60, 61
Stapp, John Paul, 20, *21*
Star Trek, 29–30, 45
Star Wars, 45
Stein, Judy, 185
Sterling, Rod, 41
Stewart, Jimmy, 35
Sullivan, Kathryn D. "Kathy"
   appearance, weight, and physical fitness of, *176*, 179–80
   The Dream is Alive filming, 166
   media coverage of, 177
   mission assignment and media coverage of, 194–95, *195*, 196–97
   mission assignments for women, 192
   qualities and physical capabilities needed for spaceflight, 190, 195
   space campers, encouragement of, 74
Sutherland, Donald, 50

technological utopianism, 83–84
television programs
   dangers and risks of spaceflight, dramatization of, 17, 18, 29
   government department involvement in production of, 9, 15–16, 20, 30
   independently produced programming, 10, 12–15, 30, 31n15
   influence of on culture and public perception, 10
   live programming, 11, 13
   masculinity and masculine ideals, representation of, 4, 22–28, 29
   network programming, 11, 13, 15, 30, 31n15
   pilot-astronauts, representation of, 48–51
   programming and the television industry, changes in, 10, 12–13, 29–30, 31n15
   rationales for and value of space exploration, 17–18, 29
   reruns and market for rebroadcasting programs, 11, 14, 30
   science and fact-based programming, 9–10, 11, 14, 15, 29–30
   scientists and scientist-astronauts, representation of, 42–43, 45–46

sets and costumes, 12, 15, *16*, 19–21, *19*, 22
spacecraft concepts and characteristics of crews, 37–39, *39*
space opera and early space television, 10–12, 38, 52n6
stereotypes introduced and reinforced in, 4, 22–28, 40–42
transition from space opera to drama and fact-based programming, 4, 9–10
Westerns, 14–15, 38
women, representation of, 23–25, 29–30, 33n51, 38
Tereshkova, Valentina, 2, 5, 116–17, 130–31, *131*, 212
Thagard, Norman E., 190
Thompson, Lea, 50
Thompson, Milton "Milt," 35–36, *36*
Thornton, William E., 54n35, 190
Time-Life, Inc, 114, 123n39
Time magazine, 65, 67, 71
Titov, Gherman, 115, 127, *131*, 145n57
To Fly, 152, 153
Tolstoy, Aleksei, 84
Tom Corbett, Space Cadet, 11–12, 15, 38
Tors, Ivan, 14, 20
Turner, Frederick Jackson, 4, 58, 61, 67–68, 69–70, 71, 73, 74
Twelve O'Clock High, 41, 53n19
The Twilight Zone, 40–41
2001: A Space Odyssey, 45

United Artists, 30
United States (US)
colonization of space and frontier settlements, 69–70
education system and competitive edge over other countries, 73–74
exceptonalism, 58, 60–62, 70
materialism and soft and indulgent living, 61–62, 65–66
national heroes, 57–58, 67–68, 76–77
reaction to human spaceflight in, 10
scientific and technological development in, 60, 70, 73–74
United States Information Agency (USIA), 125, 137, 138
United States Space Camp, 71, 74
United States space program
American prestige and success of, 28–29, 60–62, 73–76
cooperative space programs, 2, 134–35, 145n57
failures and setbacks, publication of information about, 128, 142n19
French agreement with to fly a Frenchman into orbit, 120
funding for and downsizing programs, 66–67, 68–69, 70
low Earth orbit missions, 2
openness of space program to the media, 127, 128, 134, 137–38, 141nn12–13
superpower space race for firsts, 2, 117–18
world leadership and success of, 60–61, 73–76
USSR, 140n3. See also *Soviet Life (USSR)* magazine
utopia and technological utopianism, 83–84

V-2 rockets, 16, 113
Verne, Jules, 59
Volkov, Vladislav, 142n22
von Braun, Wernher
interplanetary travel, depiction of, 59
movies, depiction of in, 48
Paris Match coverage about and articles written by, 112–14, 123n33, 123nn35–36
spacecraft concepts and characteristics of crews, 38, 52n5
space program, support for, 68
television programs, involvement in production of, 11
Voskhod spacecraft, 2
Vostok spacecraft, 2
Vysotskii, Vladimir, 83

Walker, Charles D. "Charlie," 163, 173n36, 193
Walker, David M., 179
Walt Disney television, 59, 110
Warner Bros. Pictures, 149, 172n18
Weight Watchers, 179, 180, 181
Weitekamp, Margaret, 207
West Point, 16, 31n18
White, Ed, 117, *139*, 142n22
White, Patricia, *139*
White, Robert, 40
Wolfe, Tom, 3–4, 6, 48, 204, 206–7
women astronauts
acceptance of, 6, 216, 225n30
achievements of, 197
appearance and weight of, 6, 175–77, 179–80, 192, 193, 198n7, 200n76

women astronauts (*continued*)
- careers and homelife, media attitudes toward combining, 6, 175–77, 197n3
- dating and marriage, 183–84
- domesticity and homemaking skills, 182–83
- feminism, the women's movement, and attitudes toward, 177–79
- fictional characterizations of, 6
- first woman in space, 2, 116–17, 139
- flight malfunctions and the difference between male and female astronauts, 194
- gender roles and protofeminism, 25
- hair and hairstyles for astronauts, 194
- ladies, references to astronauts at, 187–90
- media reception of, 5–6, 130–31, 143n38, 175–79, 190–97, 197n2, 197n5, 200n76
- mission assignments and media coverage of, 190–97, 200n76
- motherhood, 182–83, 184–86, *187*, 196, 199n43, 199n46
- motherhood and career balance in *Challenger Park*, 213–19, 225n30
- photograph of first women astronauts, *176*
- pilot-hero image of astronauts and, 2–3
- prohibition against female astronauts, 66
- qualities and physical capabilities needed for spaceflight, 23–25, 162, 179–82, 187–90, *188*, *189*, 197, 198–99nn30–31, 207–9
- *The Right Stuff* and representation of, 209–11
- as spacefarers, 2–3
- space shuttle astronaut recruitment and opportunities, 67, 71, 162–63, 175
- television programs and depiction of, 23–25, 29–30, 33n51, 38
- test pilot and fighter pilot opportunities for, 66

World Expo '74, 152–53, 171n13

X-15 (film), 35, 40–41, 42, 53n26
X-15 program, 19, 35–36, *37*, 40
X-20 Dyna-Soar program, 35–36, *39*

Yeager, Chuck, 50, 207
Young, John, *49*, 70, 72, 72, 161–62, 172n20
Young Astronauts Program, 71, 74
Your Future in Space, 71

Ziv Television Programs, Inc. and Frederick W. Ziv, 9, 10, 12–16, 29, 30, 31n15. See also *Men into Space*; *The Man and the Challenge*